Community School Mathematics 2A

Pat Lilburn
Pam Rawson
Peter Sullivan

Oxford University Press
Department of Education, Papua New Guinea

OXFORD UNIVERSITY PRESS AUSTRALIA

Oxford New York Toronto
Delhi Bombay Calcutta Madras Karachi
Kuala Lumpur Singapore Hong Kong Tokyo
Nairobi Dar es Salaam Cape Town
Melbourne Auckland

and associated companies in
Berlin Ibadan

OXFORD is a trade mark of Oxford University Press.

First published 1992
Reprinted 1993, 2008

ISBN 978 9980 58853 1

Written by Pat Lilburn, Pam Rawson and Peter Sullivan
Writing consultant Elsie Kinavai
Papua New Guinea project co-ordinator and writing consultant Katherine Schneider

Cover photograph by Rocky Roe Photographics
Illustrated by Juli Kent, Annie Vanston and Mary Ann Hurley
Typeset by Solo Typesetting, South Australia
Printed in Australia by Ligare
Published by Oxford University Press,
253 Normanby Road, South Melbourne, Australia and
the Department of Education, Papua New Guinea

Secretary's Message

This pupil's book is part of a new Mathematics Program designed and written for use during the early years of schooling in Papua New Guinea. The core materials consist of two pupil books called **Community School Mathematics 2A** and **Community School Mathematics 2B**. They are accompanied by the **Teacher's Resource Book 2**. They will replace the MACS series now in use.

In the Community School Mathematics Program children are taught mathematics by first using real objects. Later, the children will use pictures of objects. Finally, the children will use number symbols to represent these objects. Each school will be supplied with a set of **linking cubes** and a set of **pattern blocks** to use during mathematics lessons. You and your pupils must also collect other real objects such as seeds, nuts, shells etc. **Always allow your pupils to use real objects during their mathematics lessons if they want to. The children will decide when they do not need the help of real objects any longer**. Remember, when children use real objects, they will understand mathematics better.

This program also encourages the children to solve their own problems and make their own decisions with confidence. In order to learn these skills, the children should talk about what they are doing in every lesson. **Since learning is most effective when it has meaning and is enjoyable, allow the children to use the language that they are most comfortable with**. It is the policy of the Department of Education to encourage the use of the vernacular during the early years of schooling, where appropriate. Encourage your pupils and pay attention to what they are saying. If the children practise these skills now, they will be better able to solve problems and make decisions when they become adults.

Finally, the National Department of Education wants teachers to be **flexible** in programming and timetabling. This book will show you some ways to do this.

J. E. Tetaga OBE
Secretary for Education

Contents

20	33	50	72	71	67	22
19	28	43	21	14	13	59
18	76	25	15	64	12	70
40	17	16	44	11	60	39
68	58	7	8	9	10	78
79	6	86	62	74	83	75
77	66	5	95	81	99	42
65	98	4	96	61	54	63
2	3	100	51	32	88	24
1	23	35	47	91	52	30

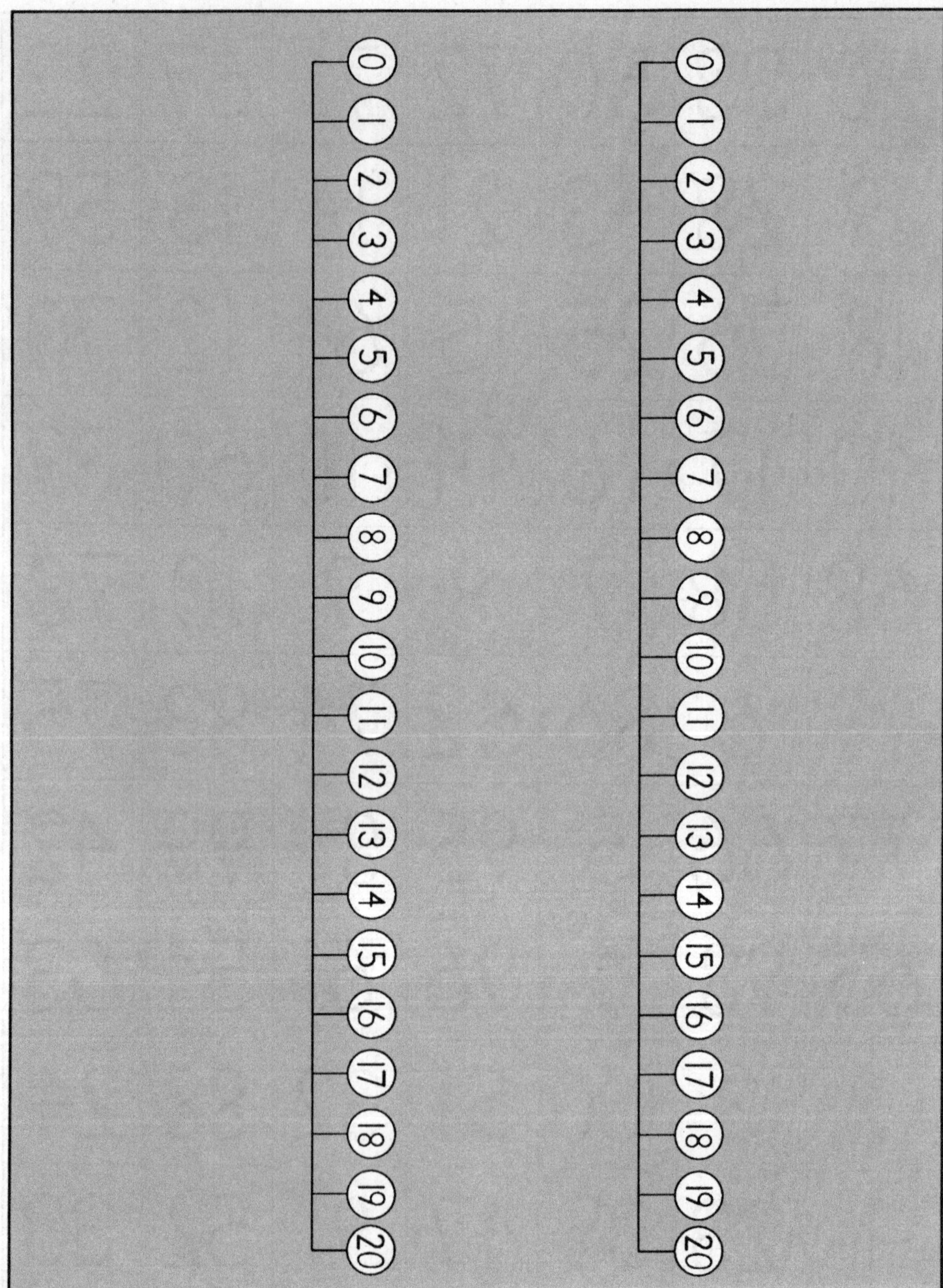
0
1
2
3
4
5
6
7
8
9
10
11
12
13
14
15
16
17
18
19
20
0
1
2
3
4
5
6
7
8
9
10
11
12
13
14
15
16
17
18
19
20

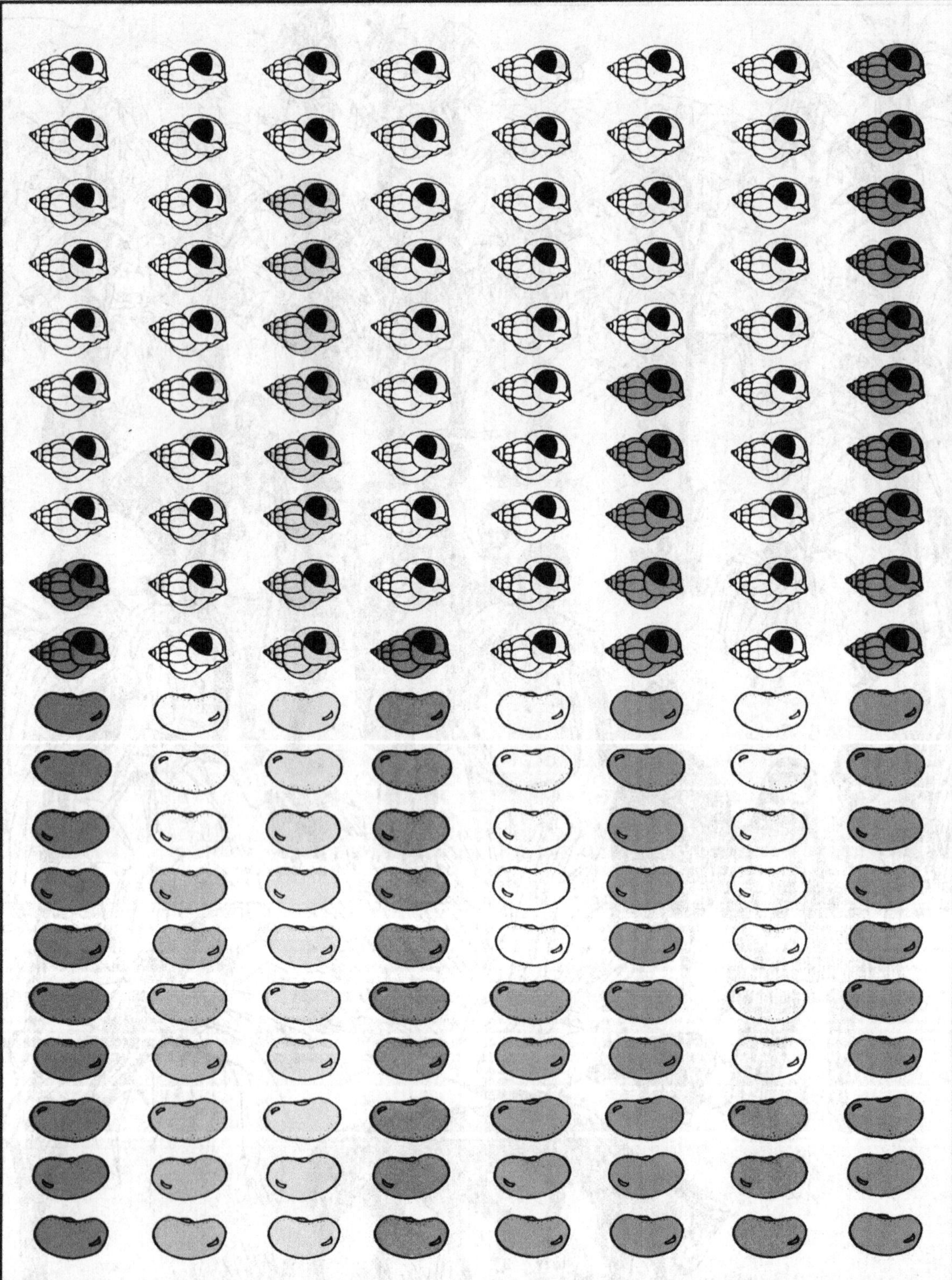

10 9 4

16

12

11 1 13

5

14

6

3

7

15 20 18

2

8 17 19

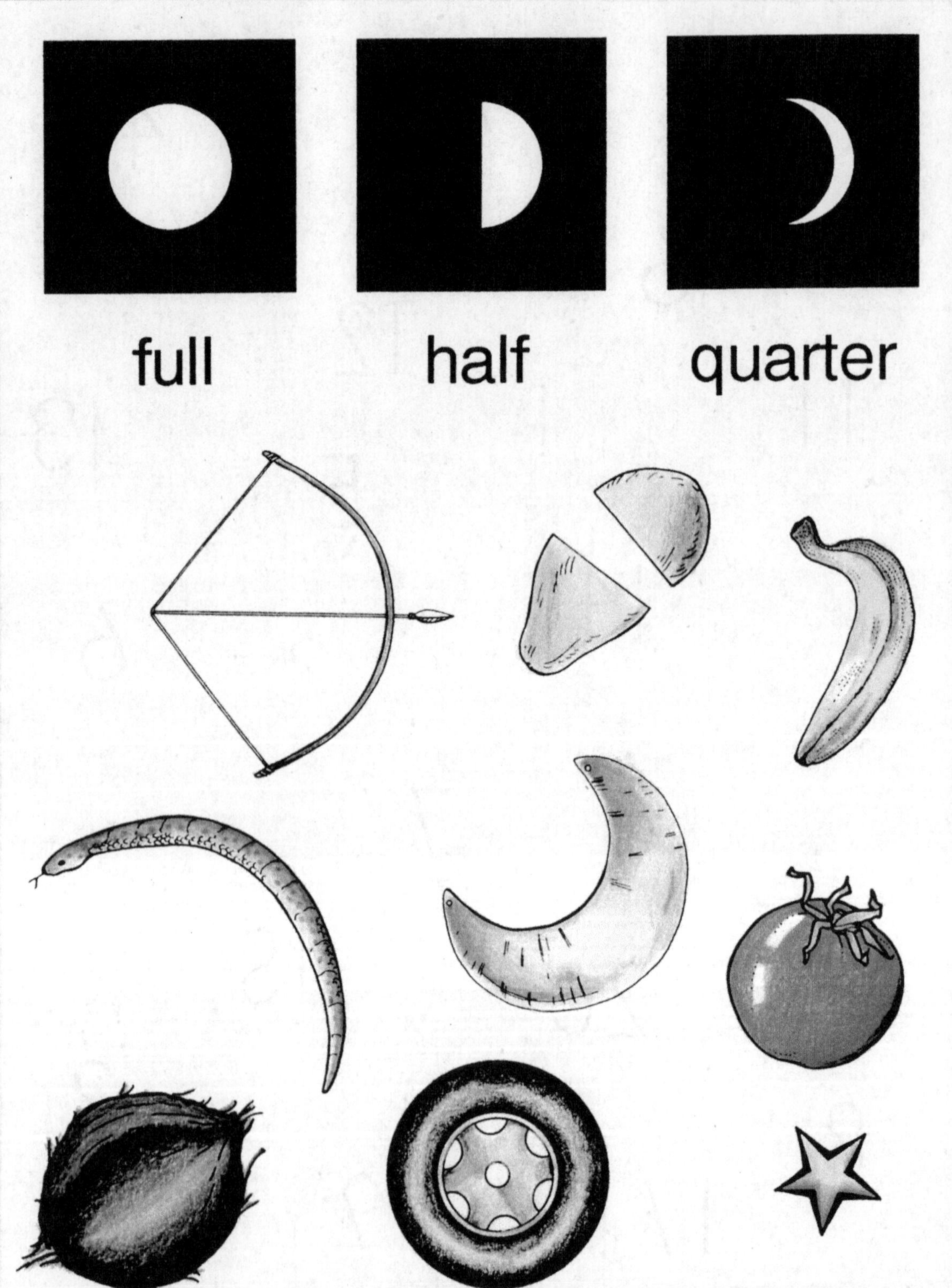
full
half
quarter

What do the stars look like?

A

B

C

D

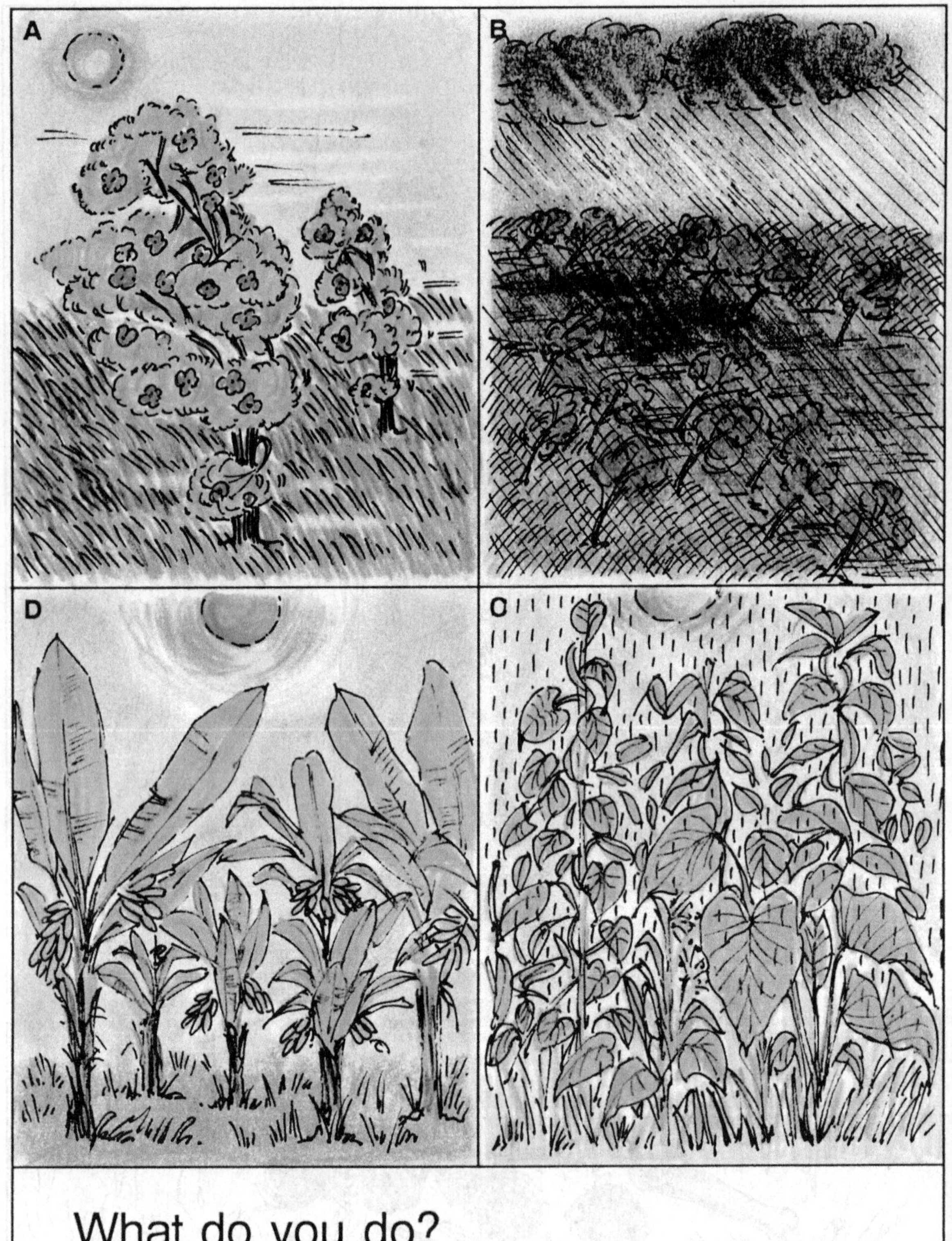

What do you do?

Start here

0	1	2	3	4	5	6	7	8	9
10	11		13	14	15		17	18	19
20	21	22	23		25	26	27		29
30			33	34	35	36		38	39
40	41	42		44	45		47	48	49
50	51	52	53			56	57	58	59
60	61		63	64	65	66		68	
70	71	72	73		75	76	77	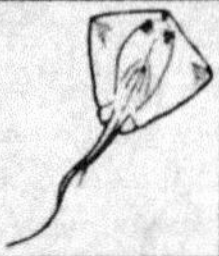	79
80		82		84	85		87	88	89
90	91		93	94	95	96		98	99
100	You're finished!								

STRAWBERRY

Short or long?

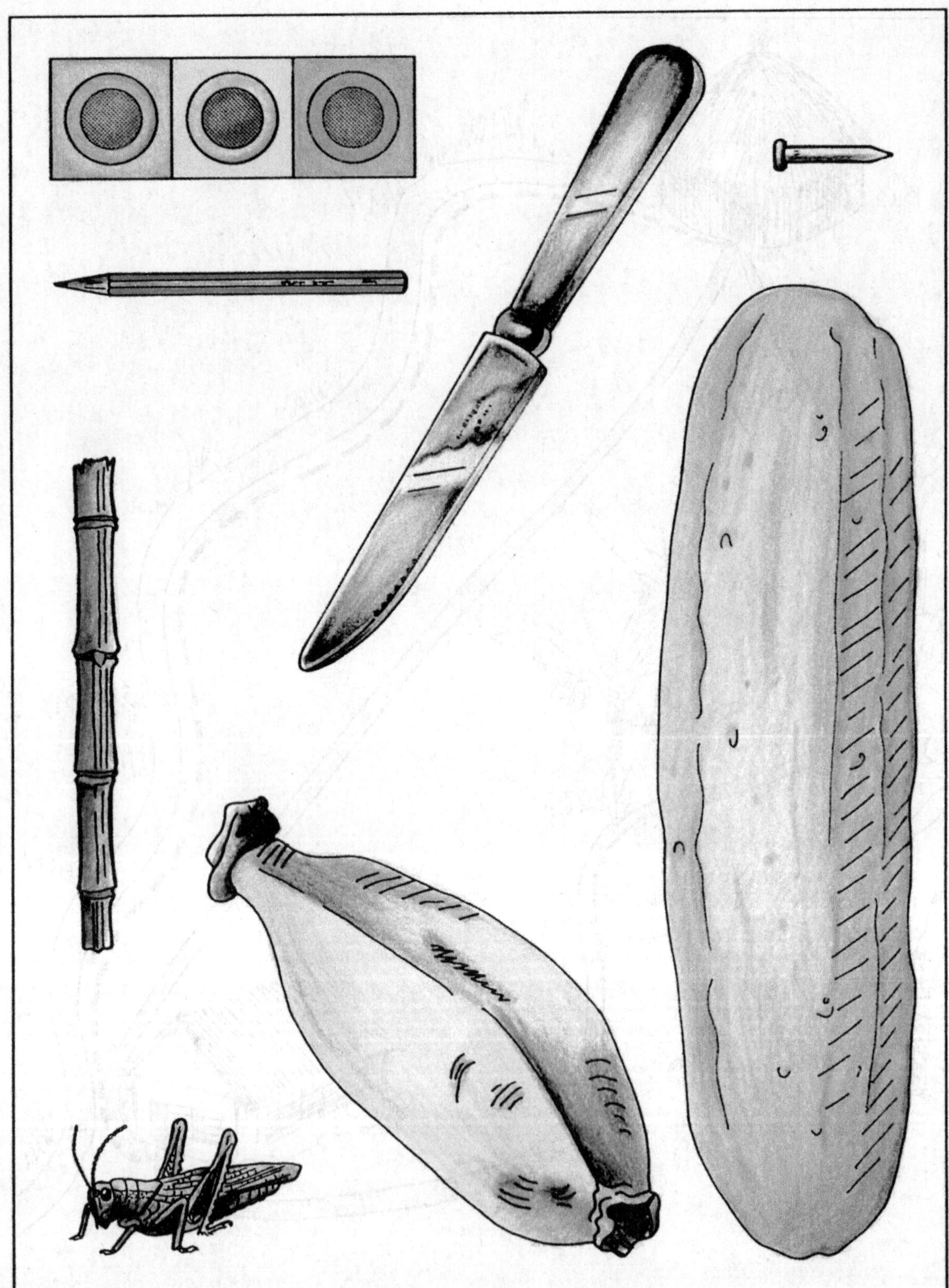

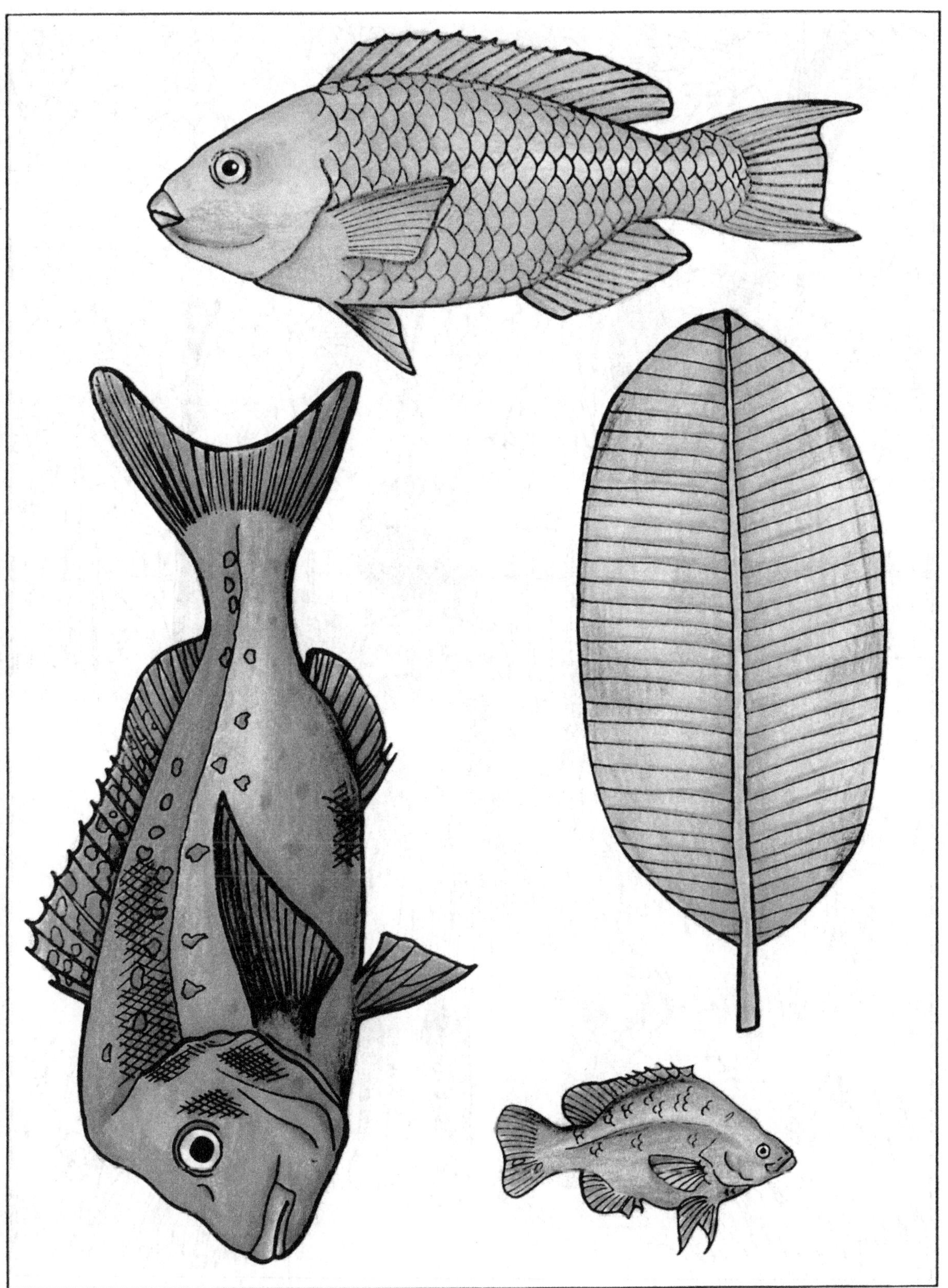

How many?

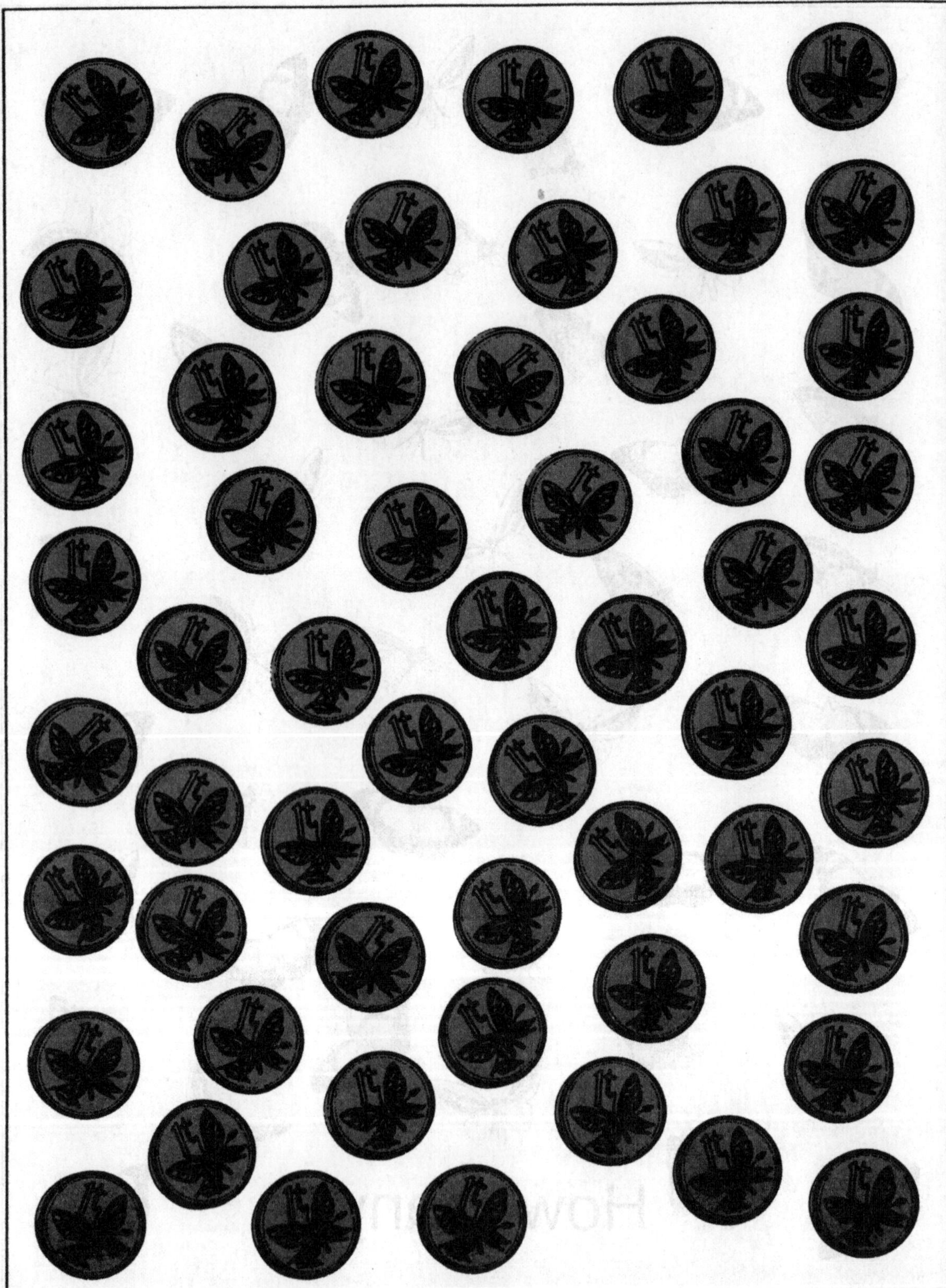

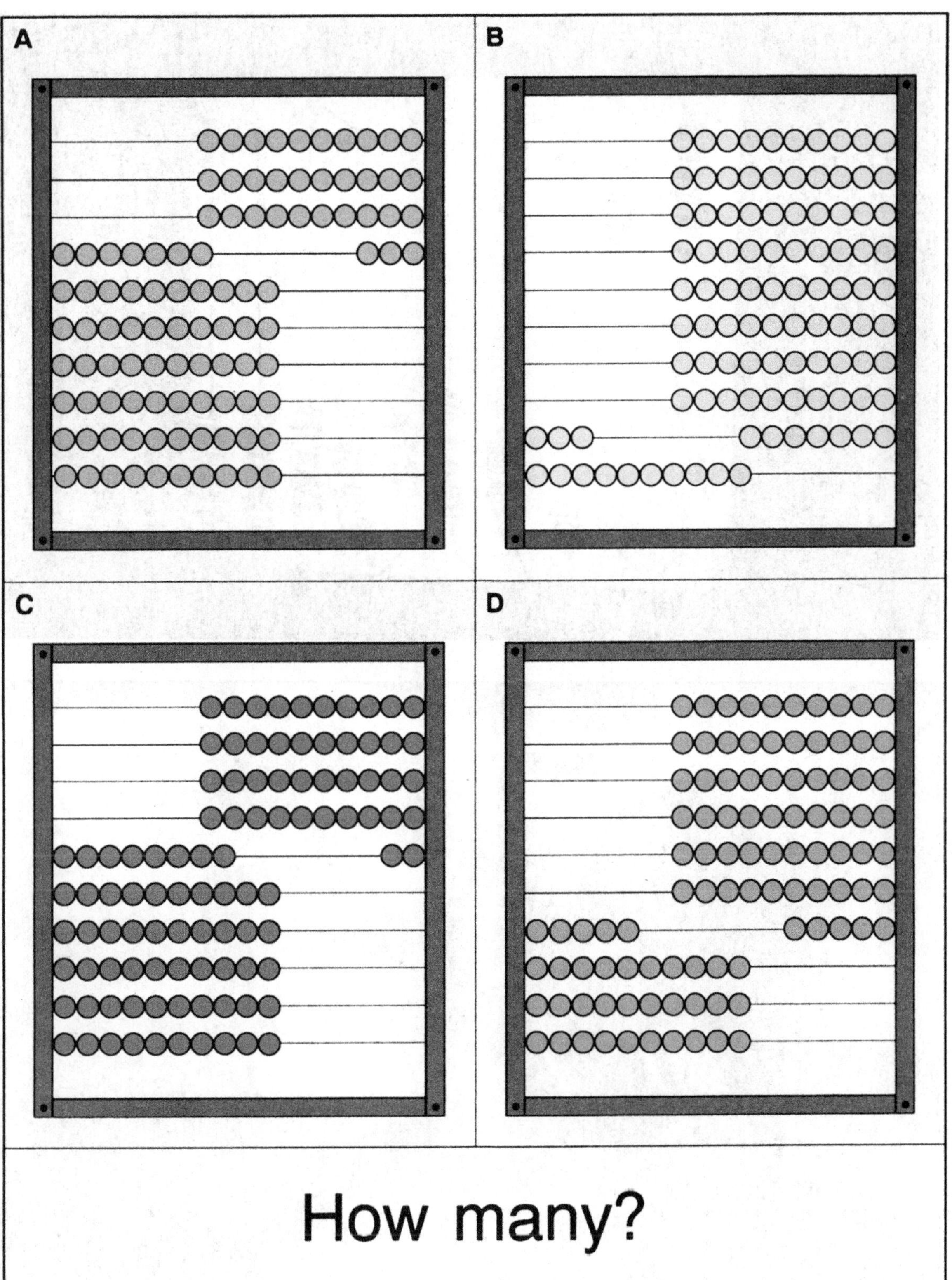

How many?

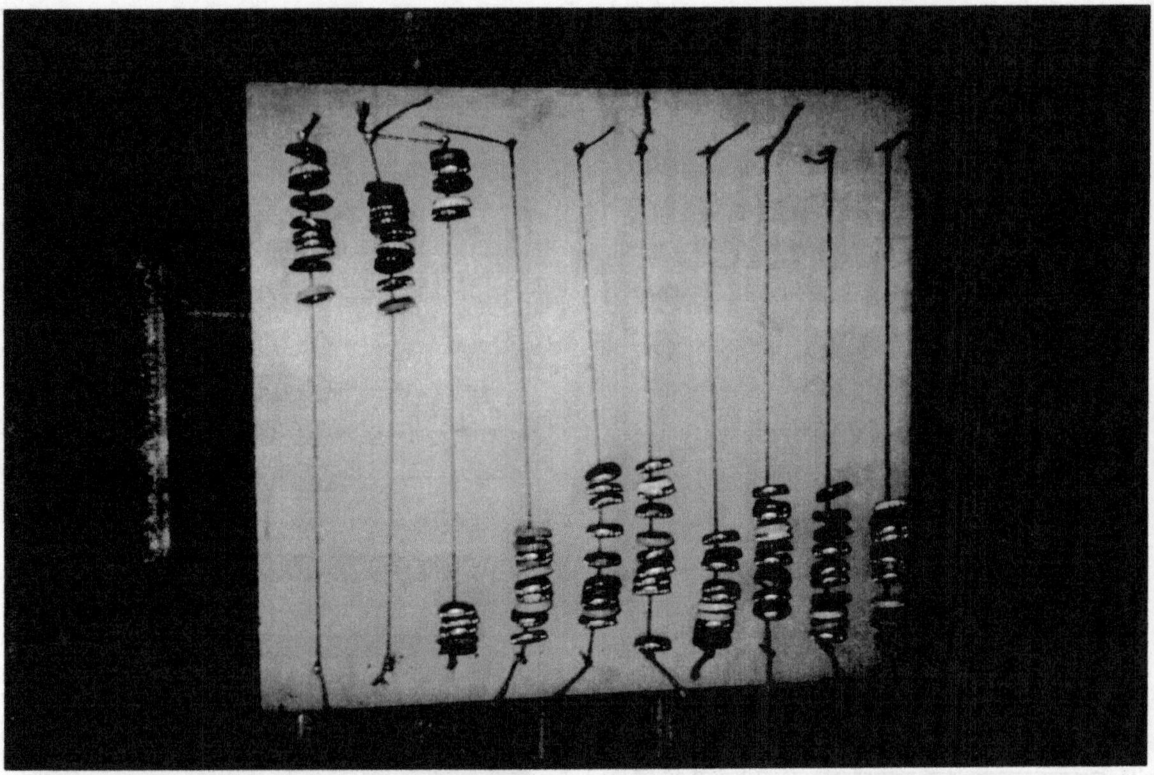

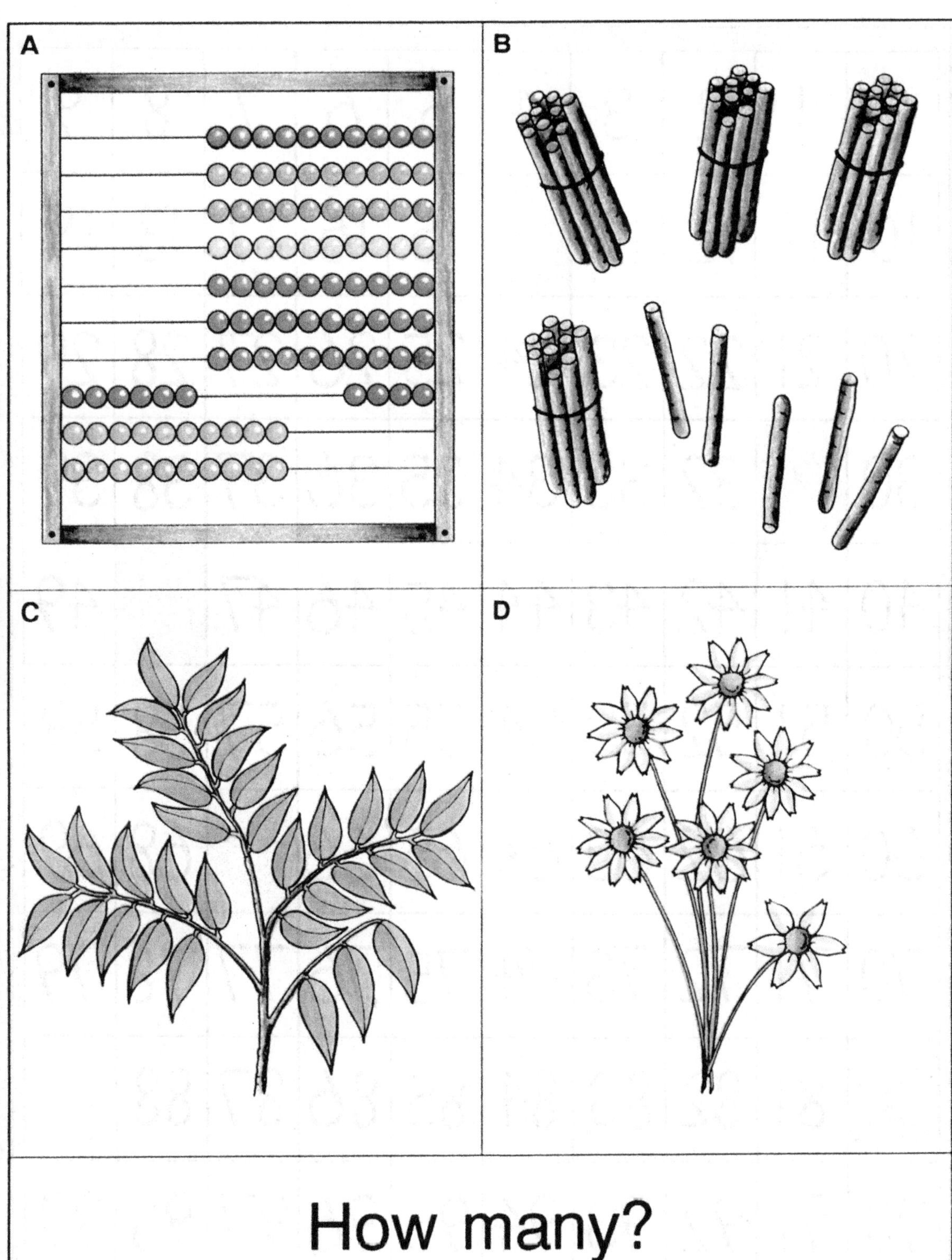

How many?

0	1	2	3	4	5	6	7	8	9
10	11	12	13		15	16	17	18	19
20	21	22	23	24	25	26	27	28	29
30		32	33	34	35	36	37	38	39
40	41	42	43	44	45	46	47		49
50	51	52	53	54	55	56	57	58	59
60	61	62	63	64	65	66		68	69
70	71	72	73	74	75	76	77	78	79
	81	82	83	84	85	86	87	88	
90	91	92	93	94	95	96	97	98	99

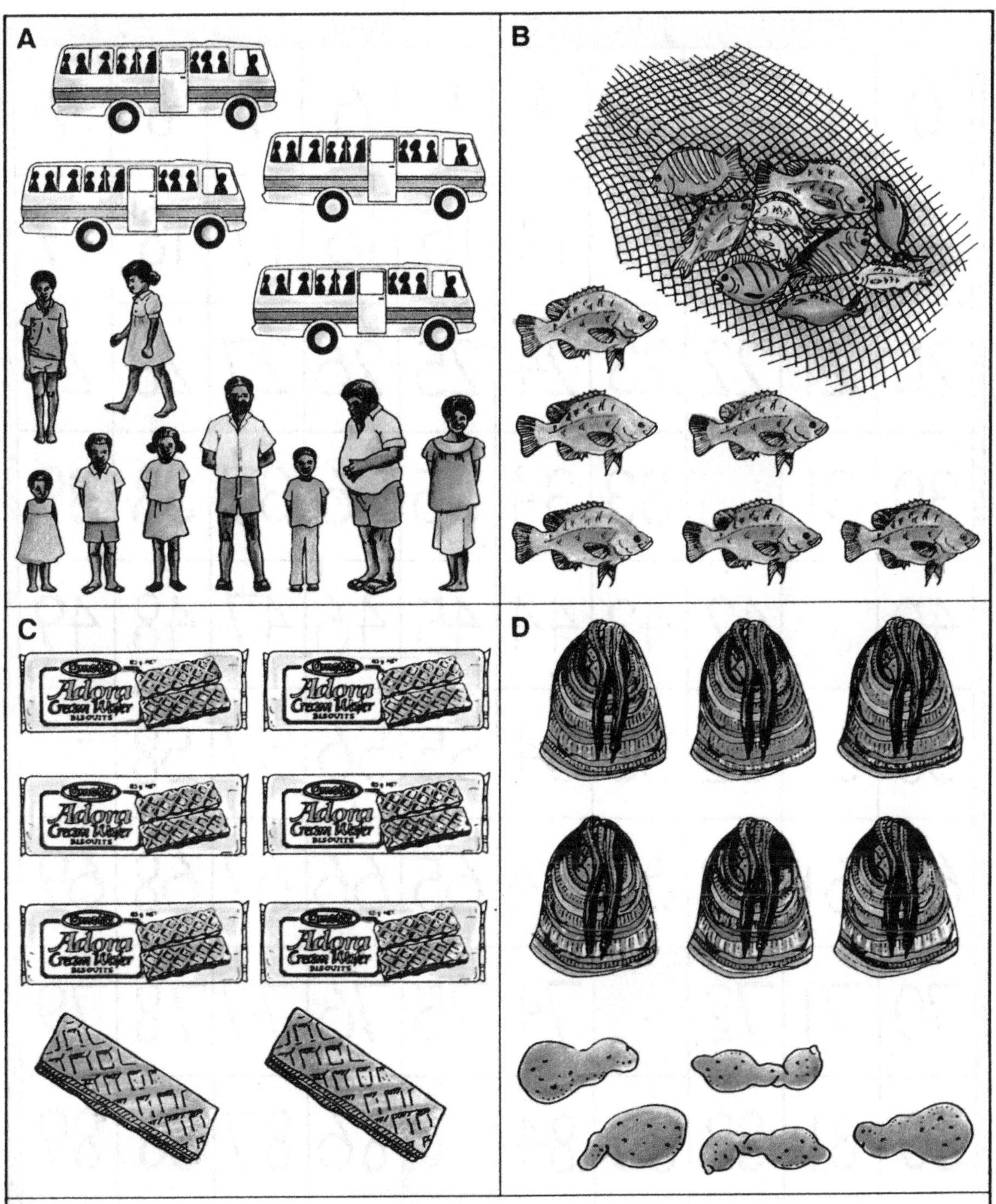

How many?

0	1	2	3	4	5	6	7	8	9
10	11	12	13	14	15	16	17	18	19
20	21	22	23	24	25	26	27	28	29
30	31		33	34	35	36	37	38	39
40		42	43	44	45	46	47	48	49
50	51	52	53	54	55	56	57	58	
60	61	62	63	64	65	66	67	68	69
70	71	72		74	75	76	77	78	79
80	81	82	83	84		86	87	88	89
90	91	92	93	94	95	96	97	98	99

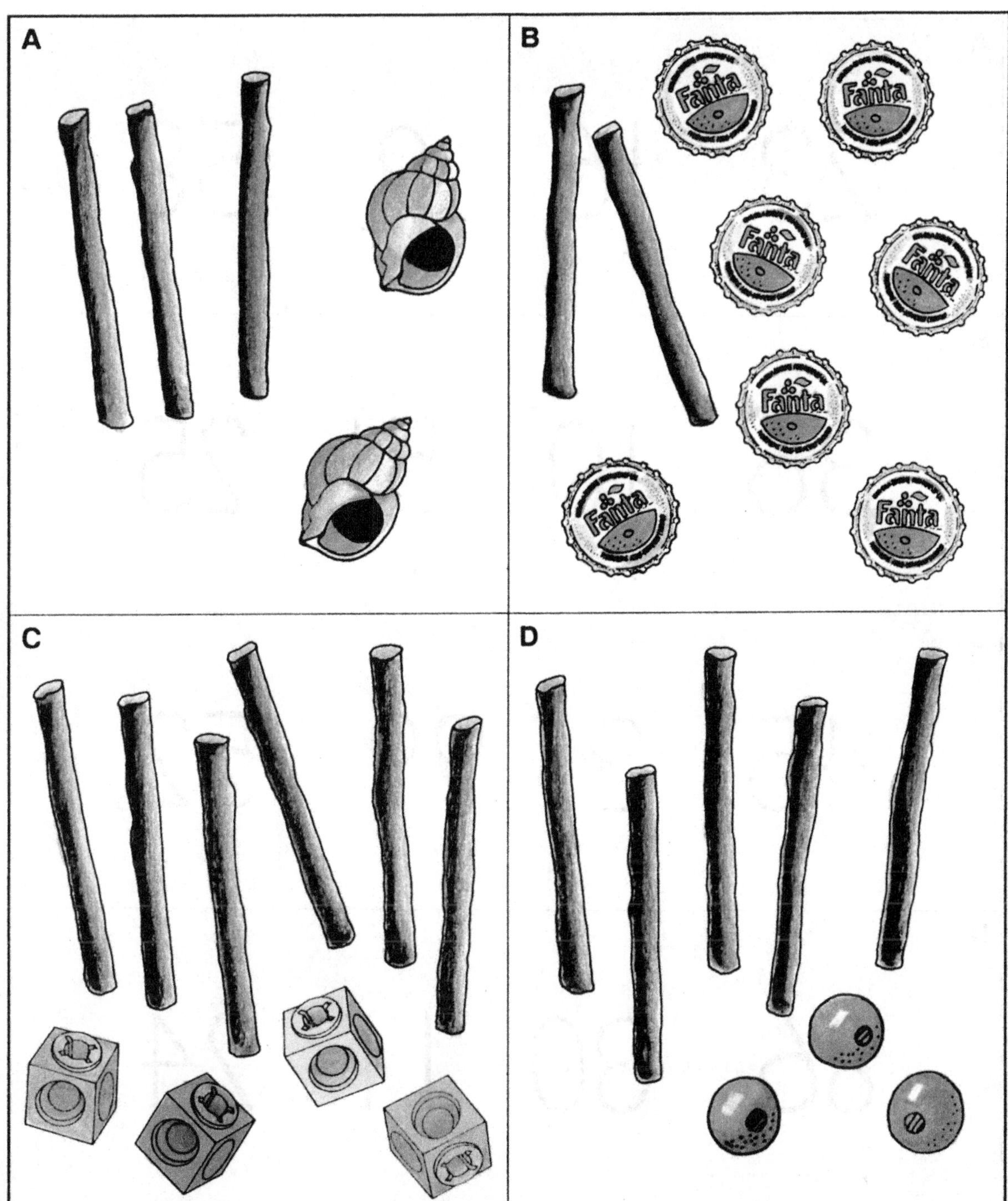

How many?

A	23	16	9	58
B	36	10	51	25
C	45	3	99	52
D	86	30	11	24

Put the numbers in order.

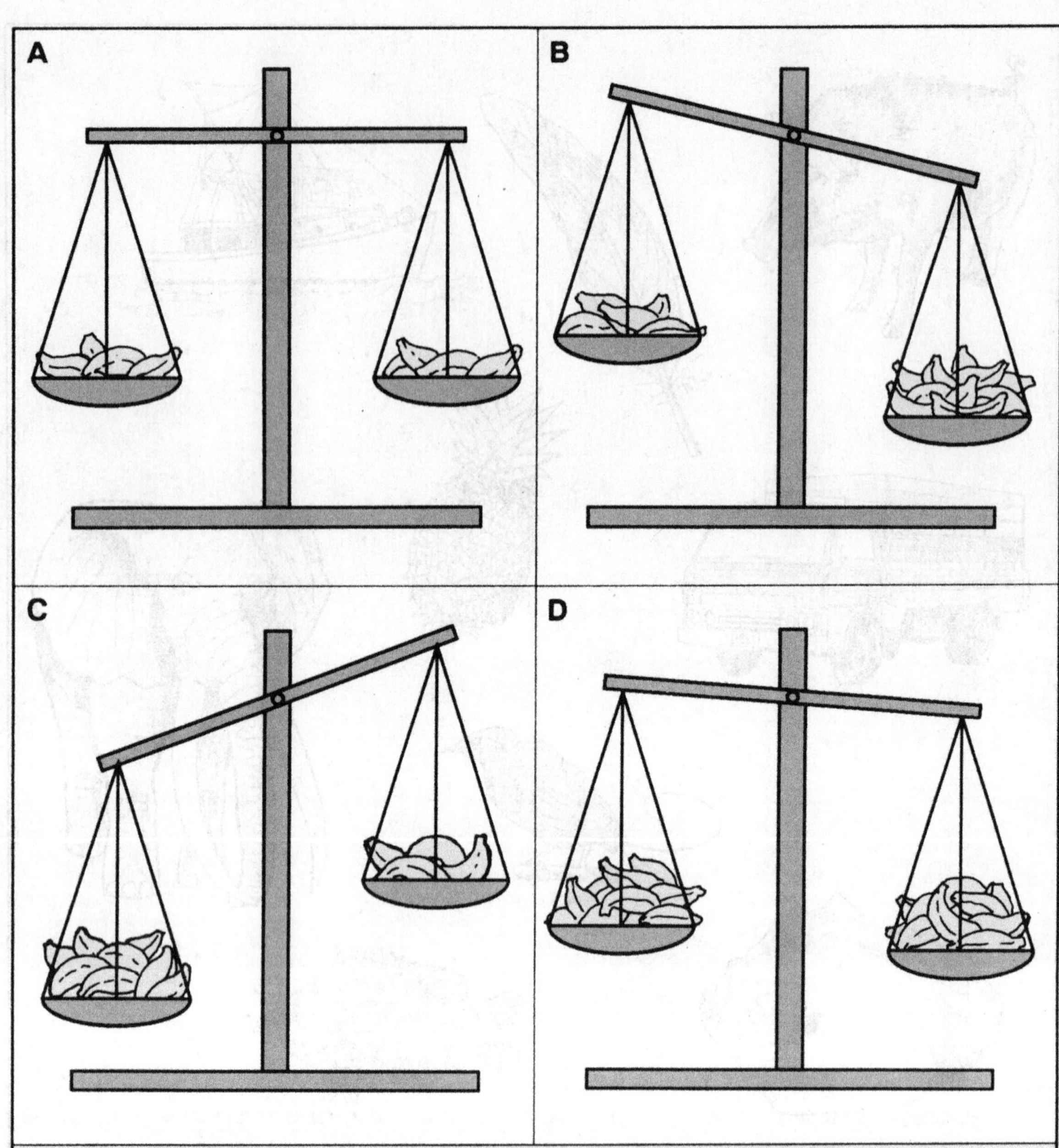

Which is lighter?
Which is heavier?

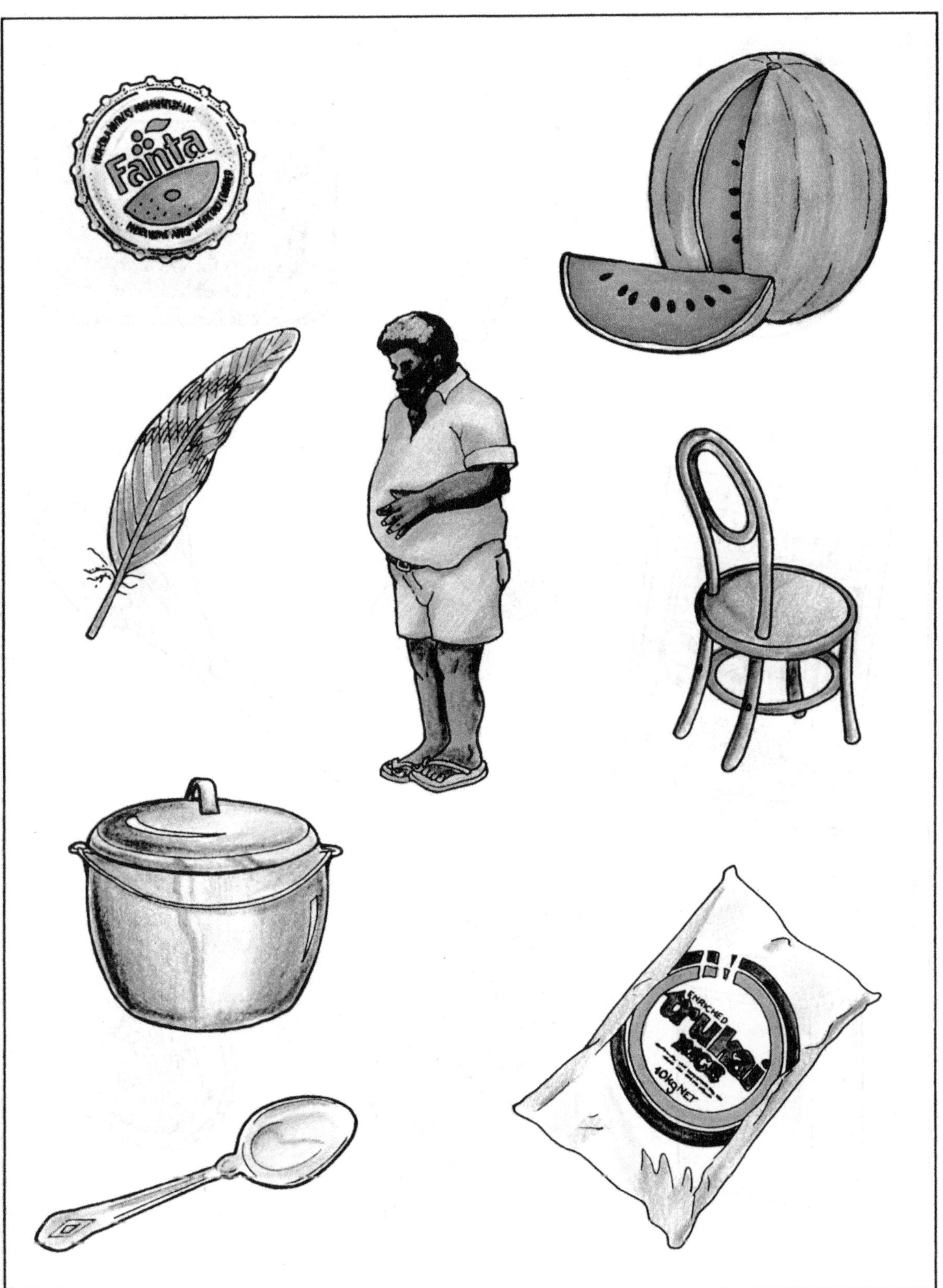
Fanta
ENRICHED
trukai
40kg NET

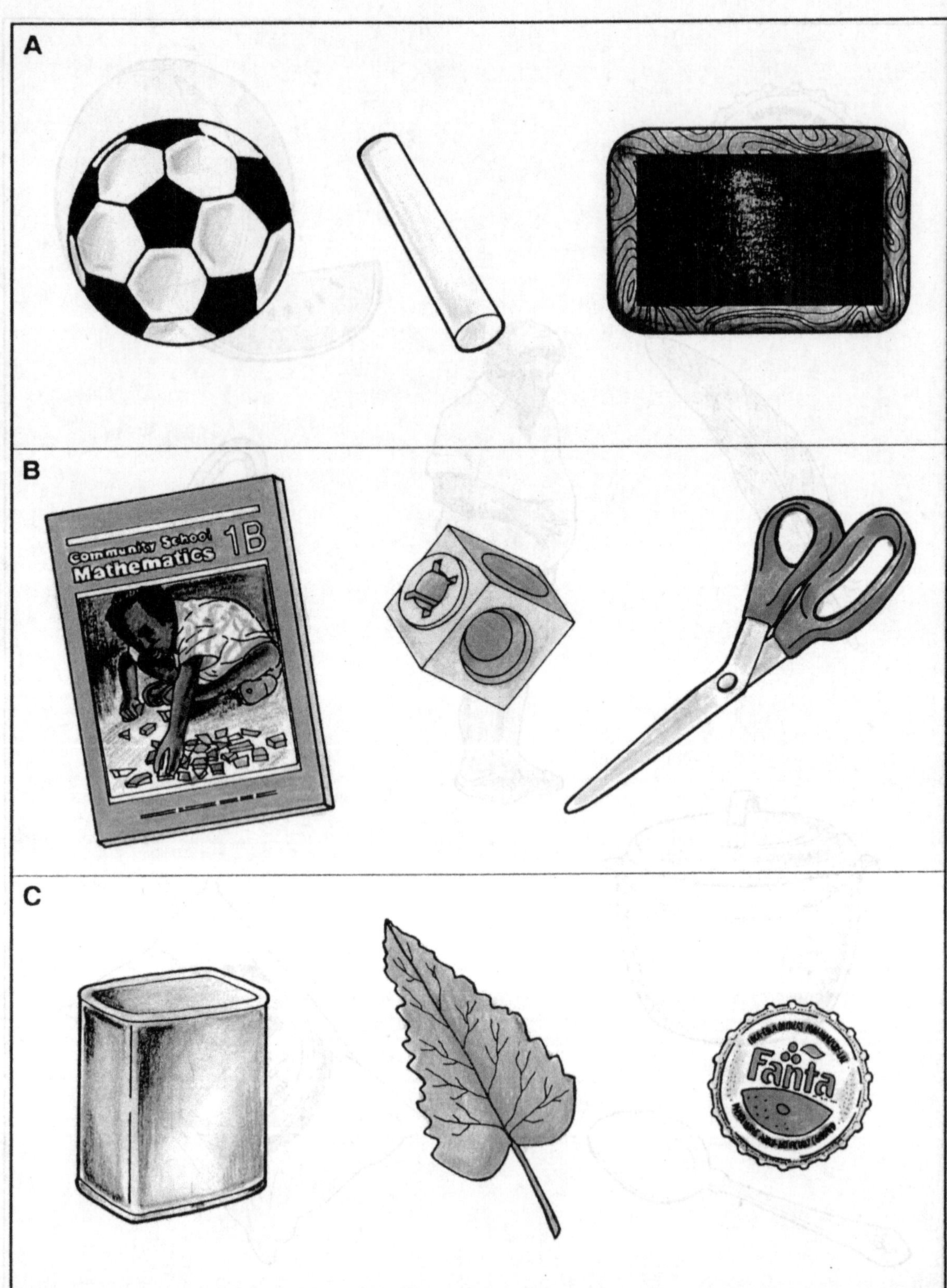
A
B
Community School
Mathematics 1B
C
Fanta

A

3 tens + 4 units

B

2 tens + 7 units

C

5 tens + 3 units

D

4 tens + 8 units

Make these numbers.

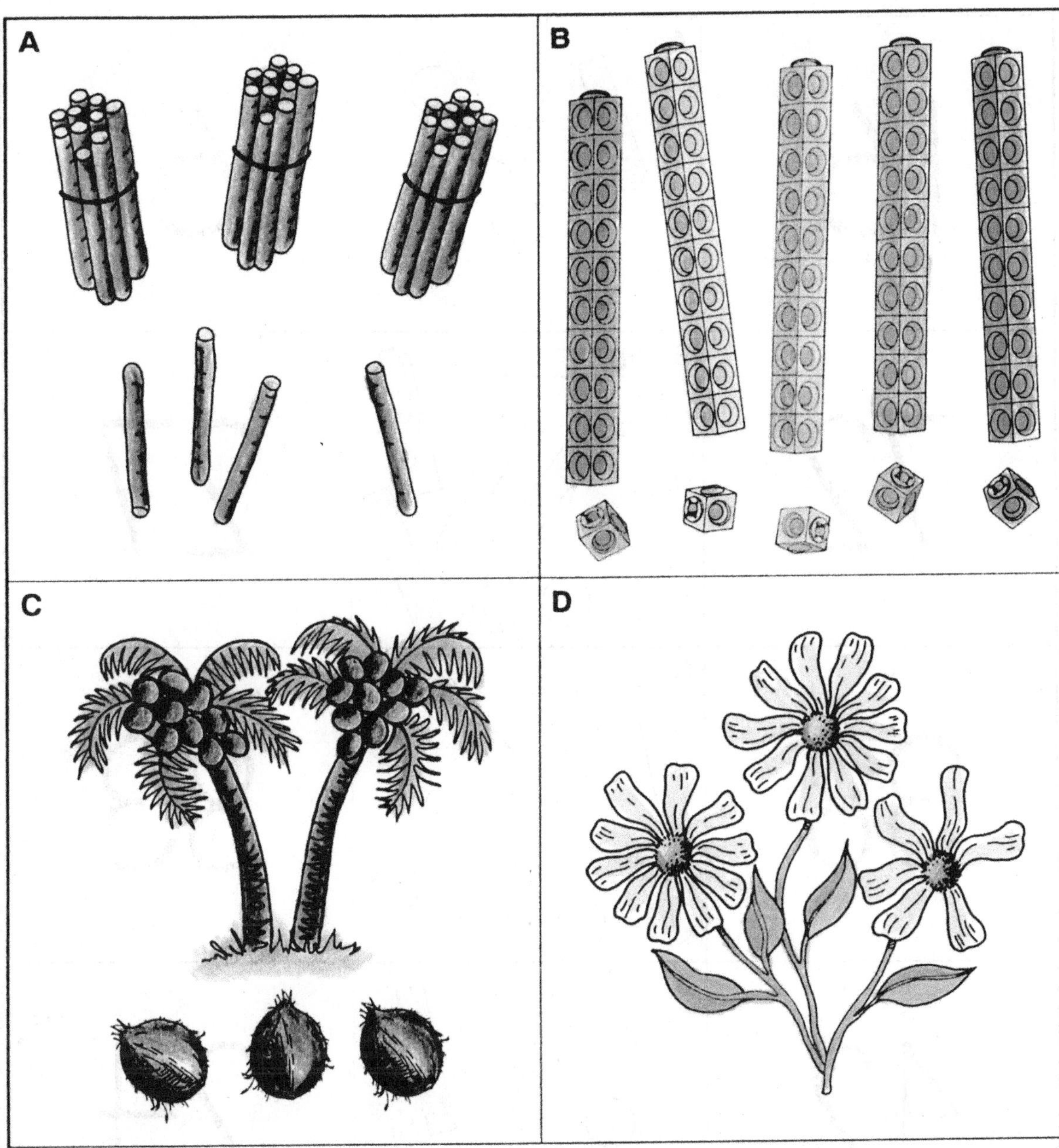

How many tens?

How many units?

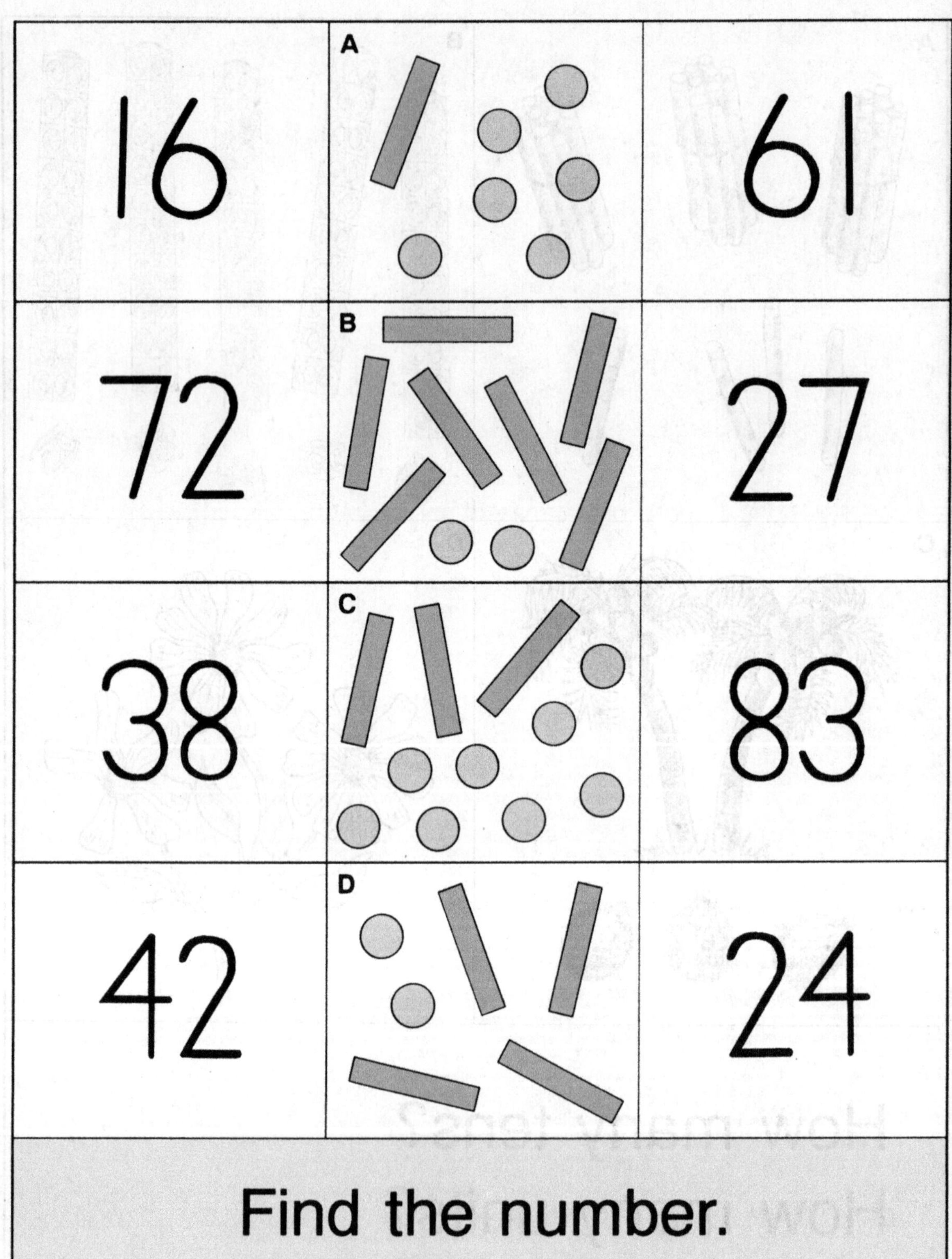
16
A
61
72
B
27
38
C
83
42
D
24
Find the number.

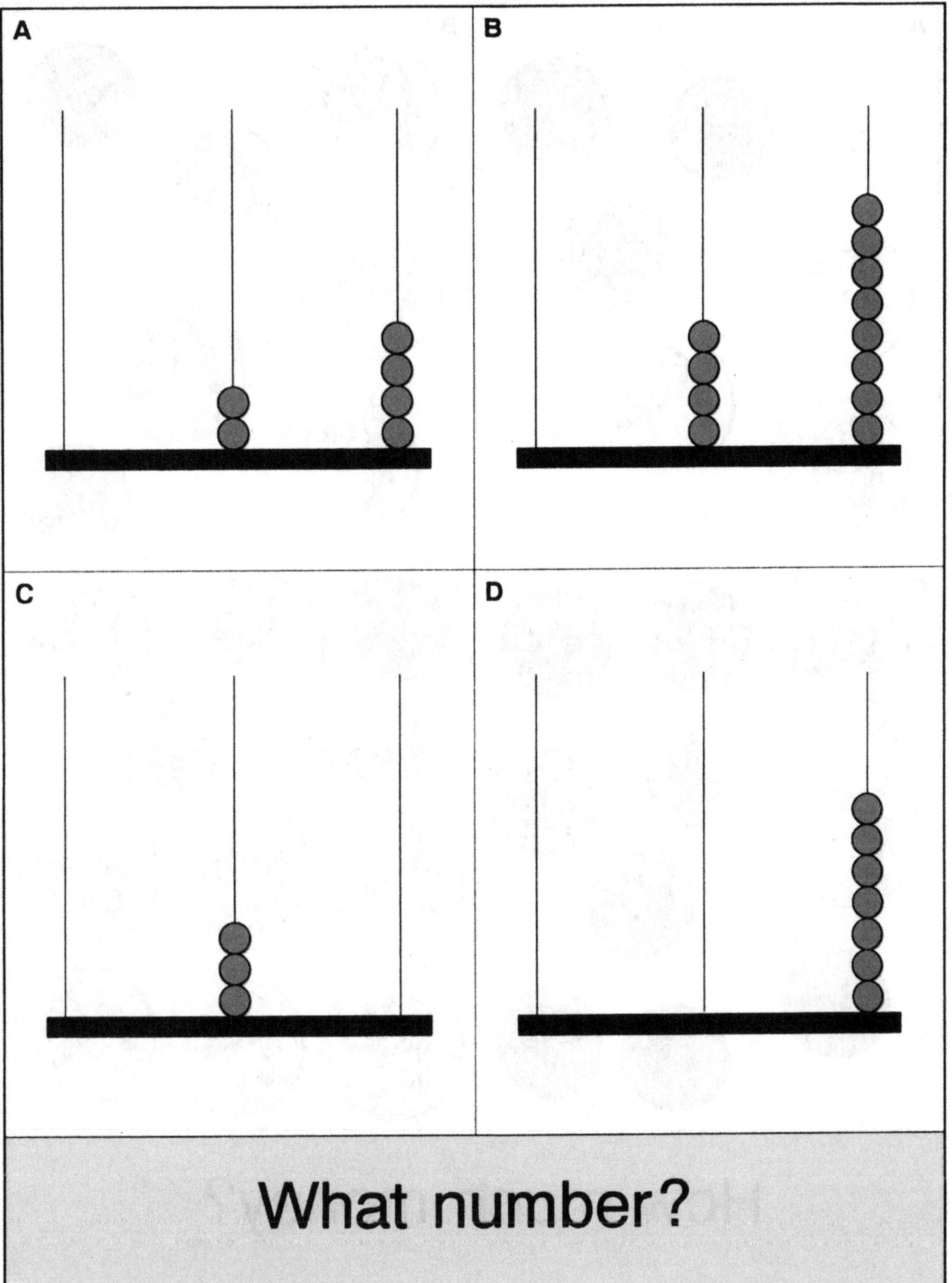

What number?

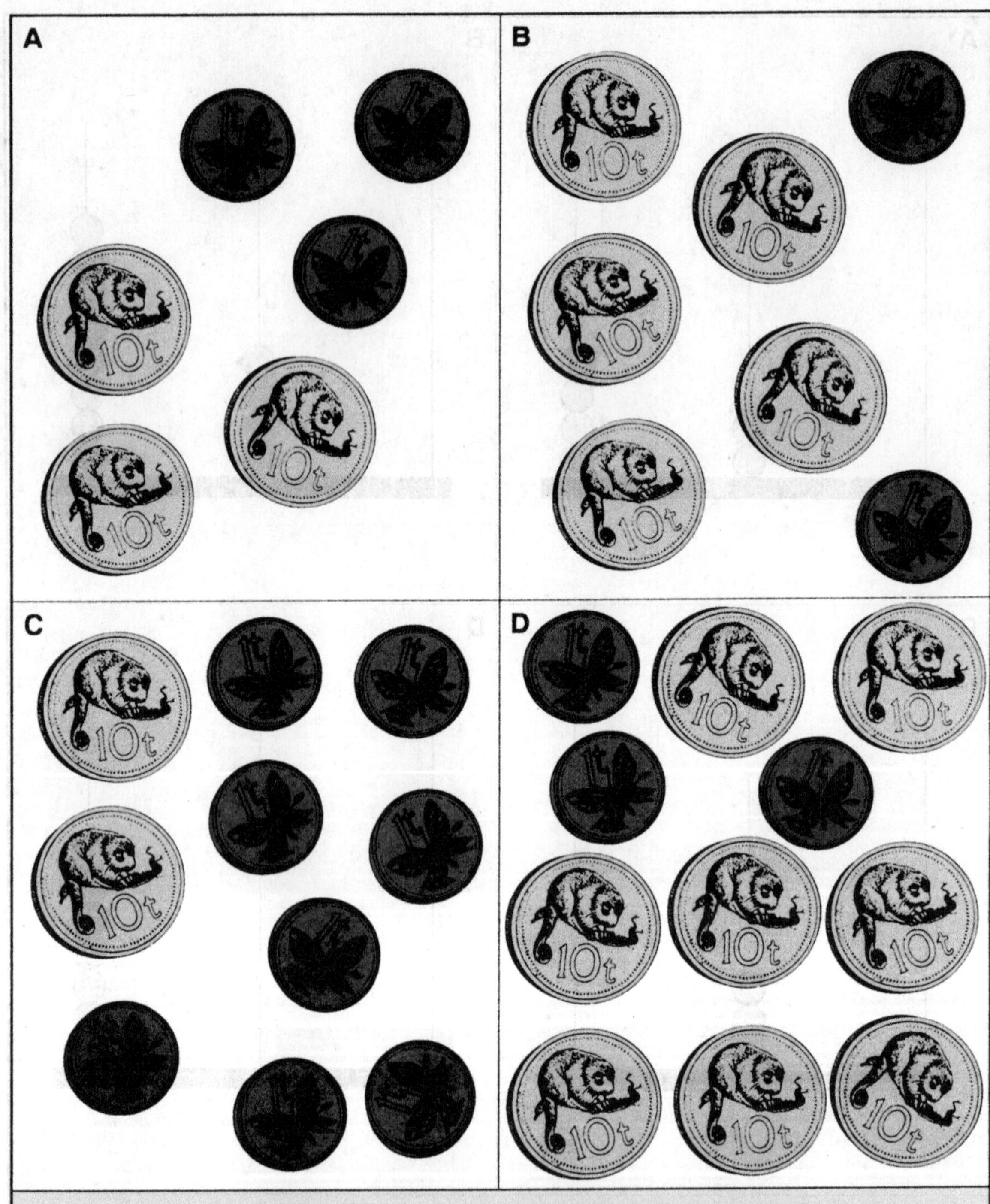

How much money?

Easter in Papua New Guinea.

Put the pictures in order.

A

B

A
B

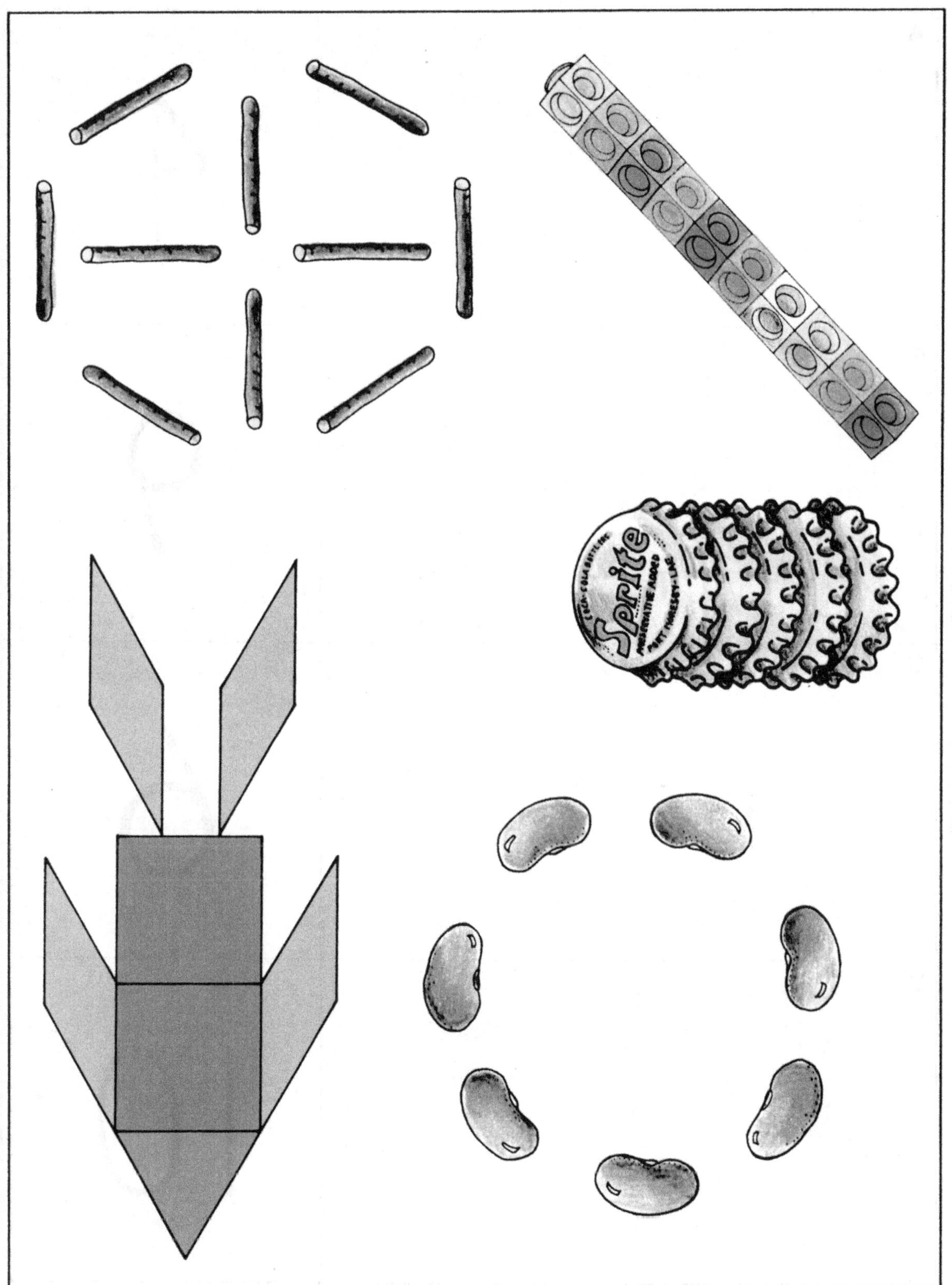
Sprite

A	4
B	9
C	6
D	10

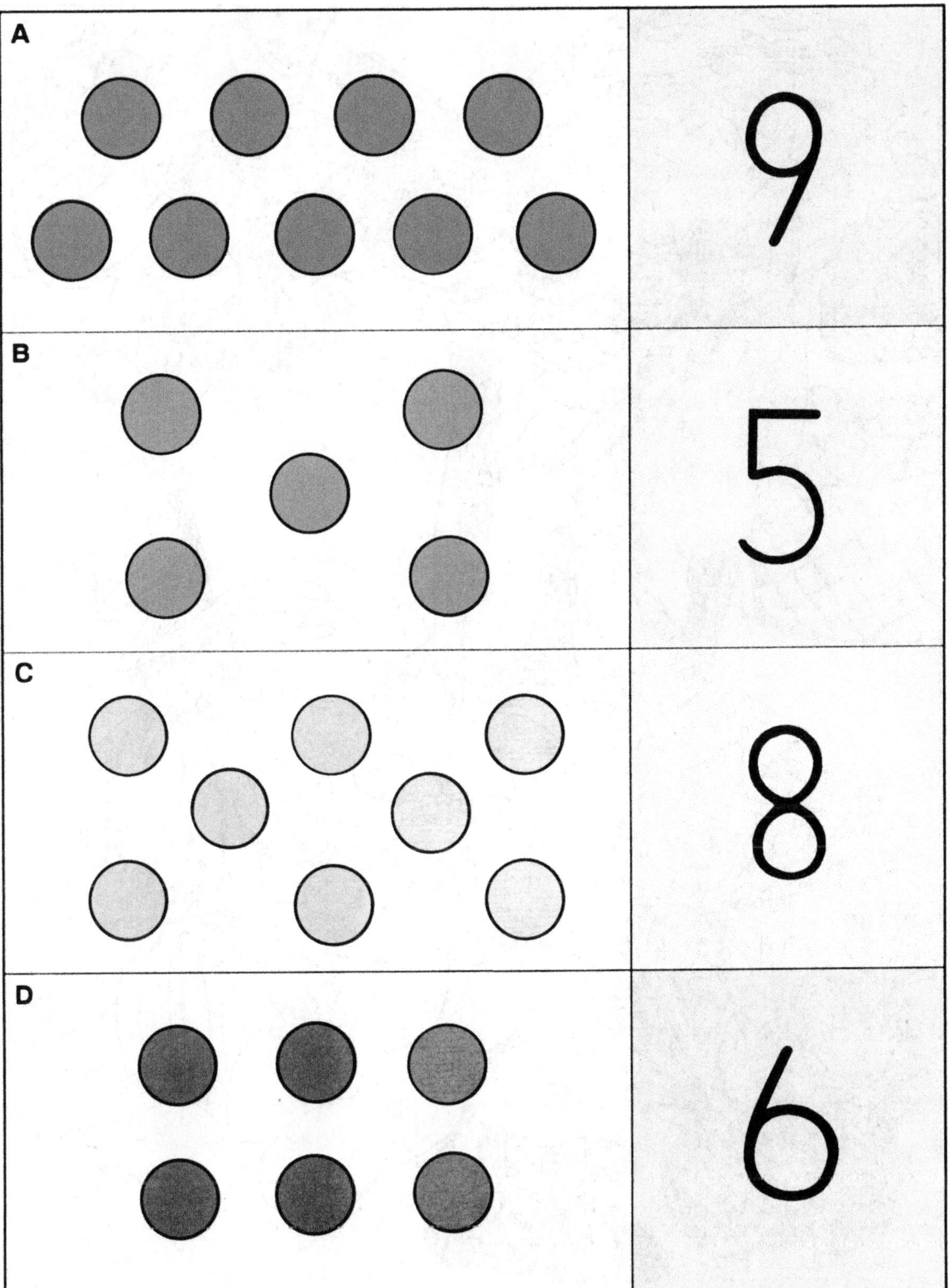
A
9
B
5
C
8
D
6

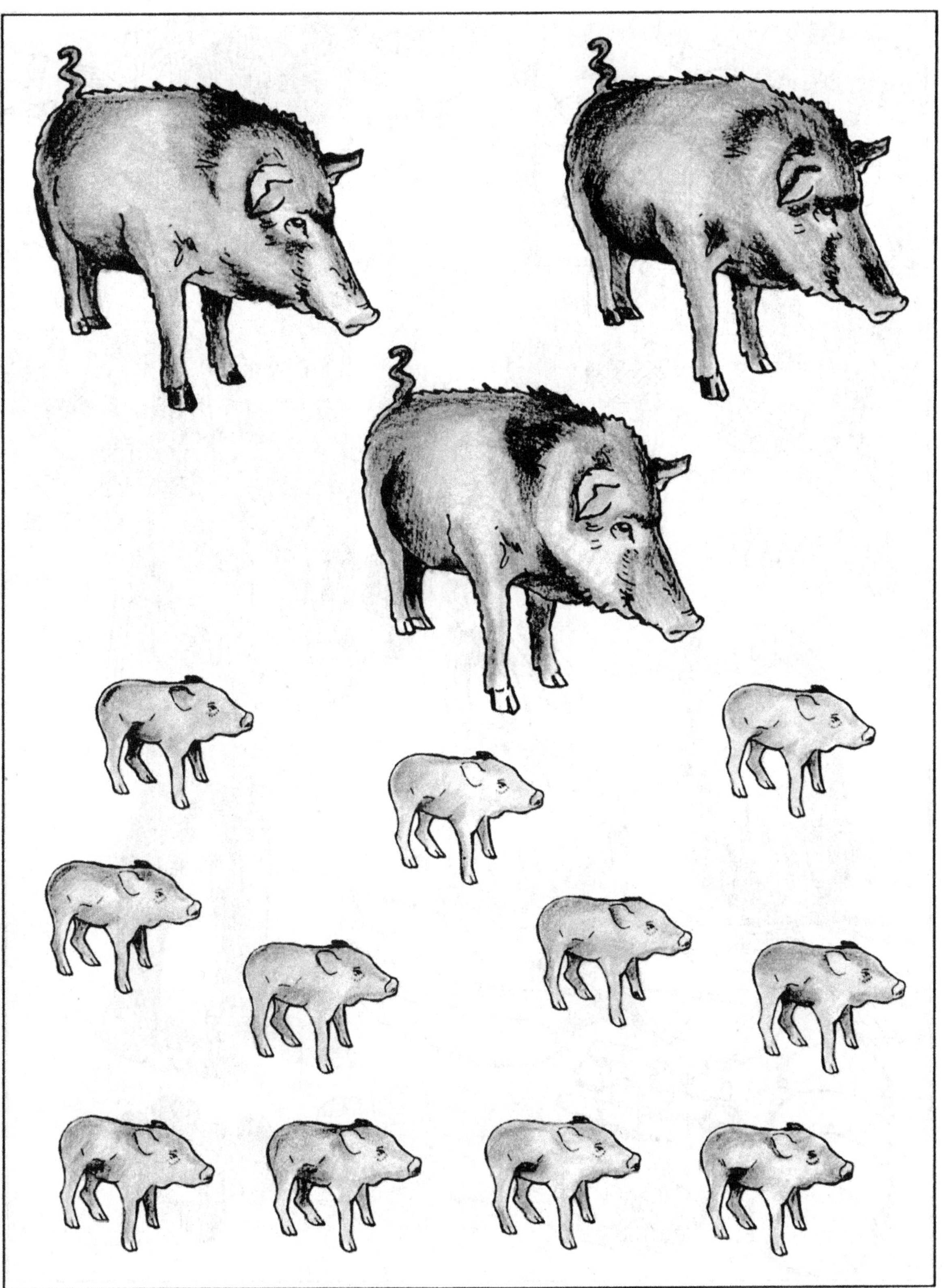

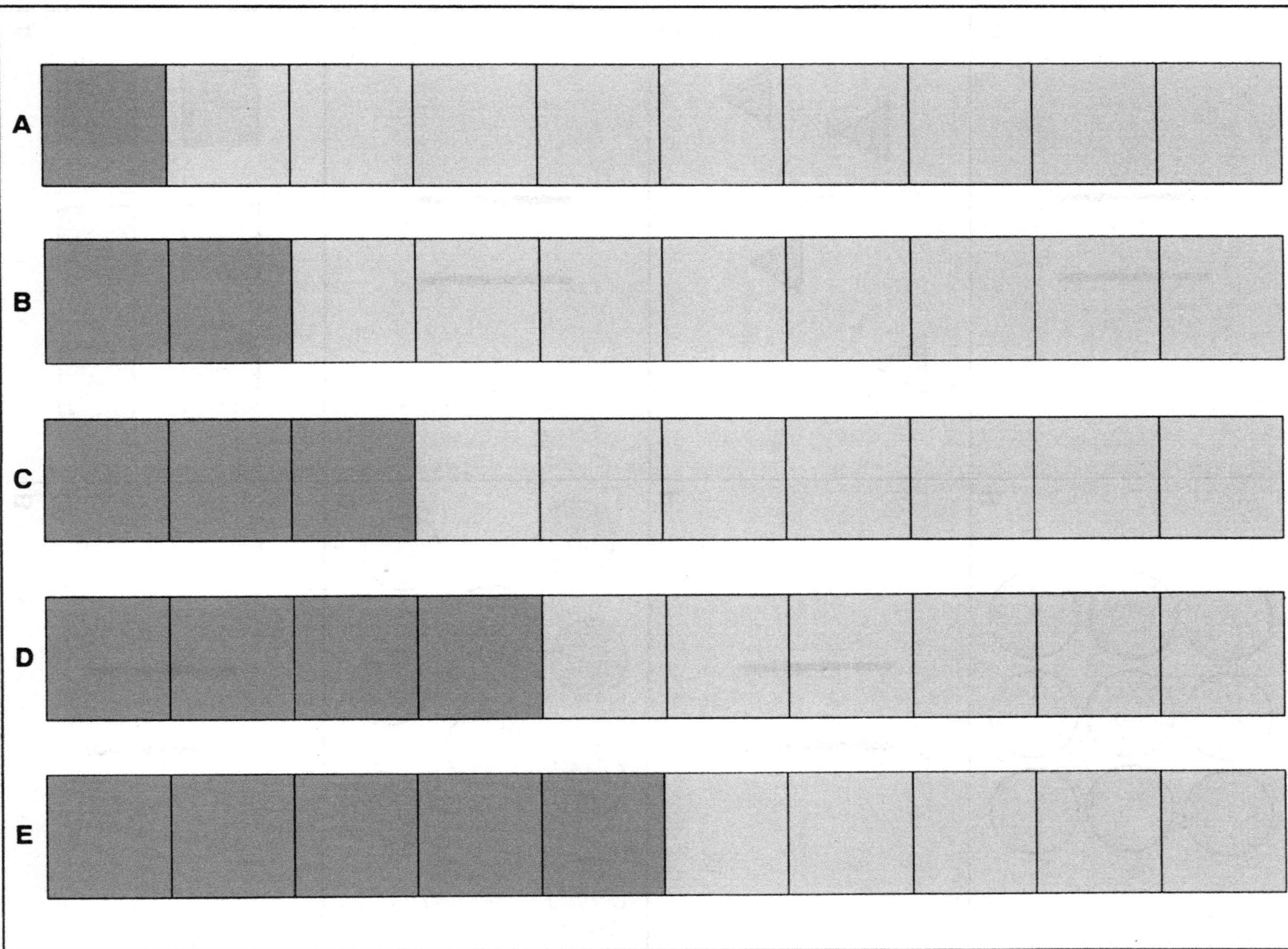
A
B
C
D
E

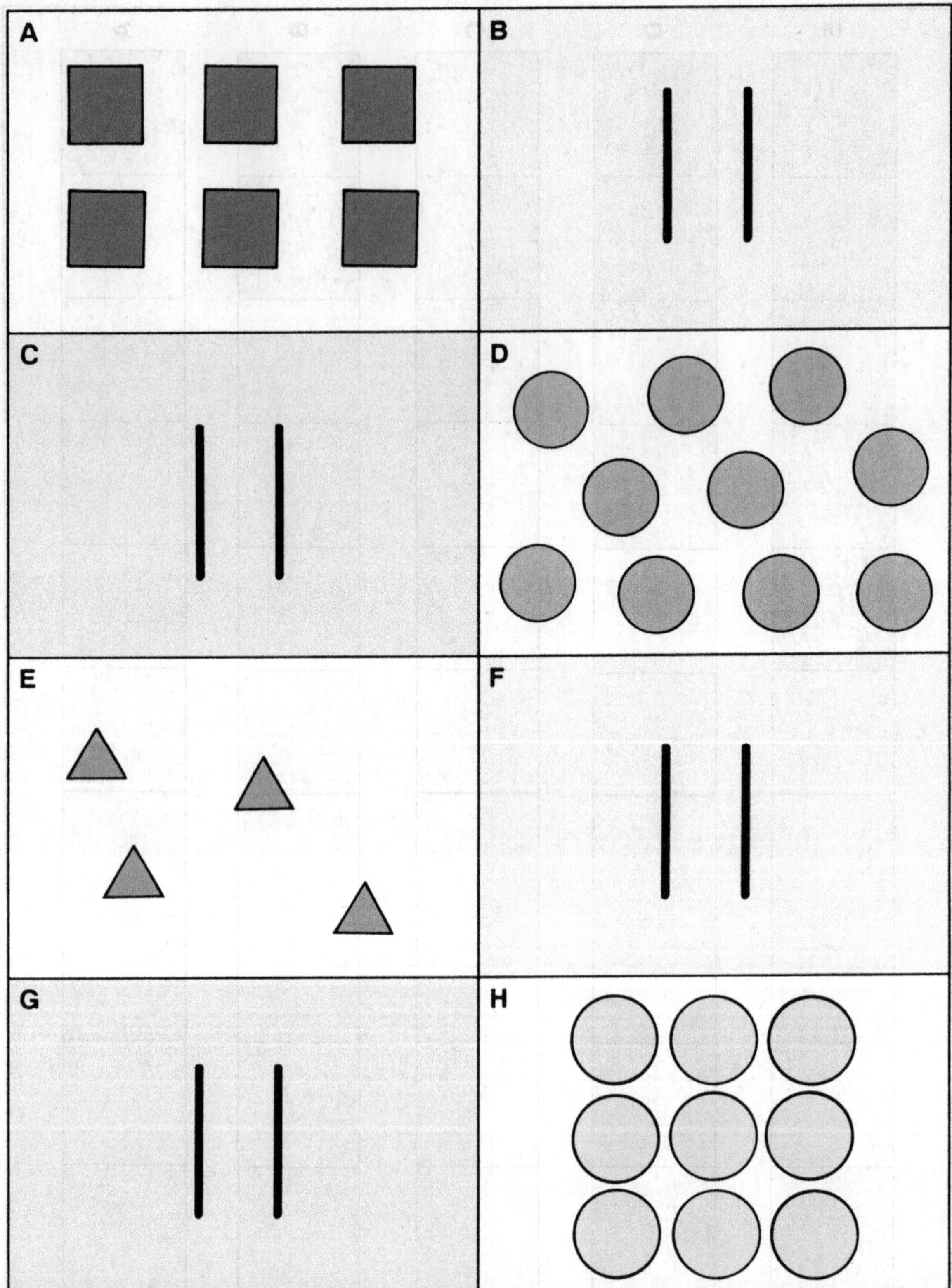
A
B
C
D
E
F
G
H

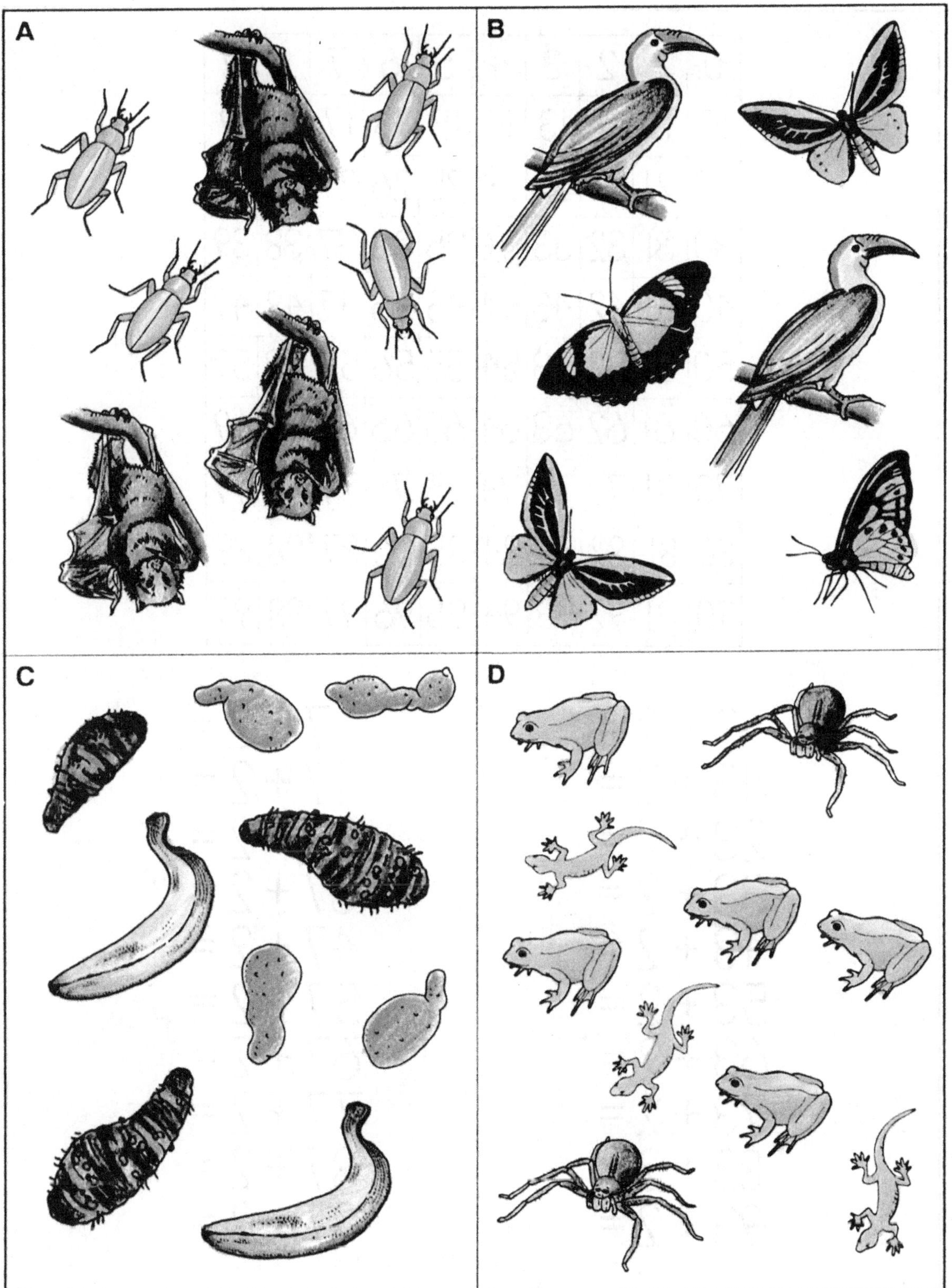
A
B
C
D

0	1	2	3	4	5	6	7	8	9
10	11	12	13	14	15	16	17	18	19
20	21	22	23	24	25	26	27	28	29
30	31	32	33	34	35	36	37	38	39
40	41	42	43	44	45	46	47	48	49
50	51	52	53	54	55	56	57	58	59
60	61	62	63	64	65	66	67	68	69
70	71	72	73	74	75	76	77	78	79
80	81	82	83	84	85	86	87	88	89
90	91	92	93	94	95	96	97	98	99

3 + 2 =
13 + 2 =
23 + 2 =
33 + 2 =
43 + 2 =
53 + 2 =
63 + 2 =
73 + 2 =
83 + 2 =
93 + 2 =

7 + 2 =
17 + 2 =
27 + 2 =
37 + 2 =
47 + 2 =
57 + 2 =
67 + 2 =
77 + 2 =
87 + 2 =
97 + 2 =

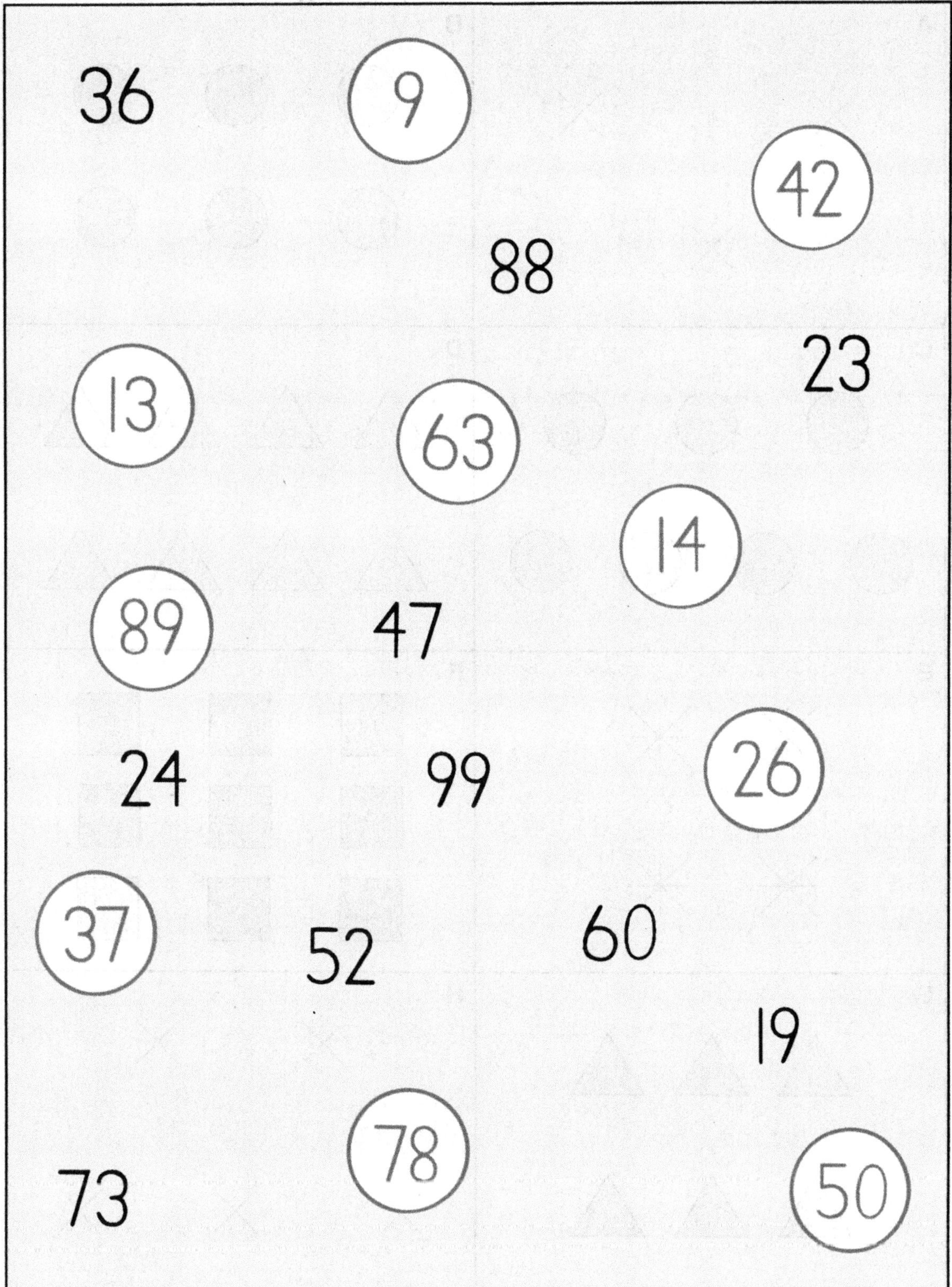
36
9
42
88
23
13
63
14
89
47
24
99
26
37
52
60
19
78
73
50

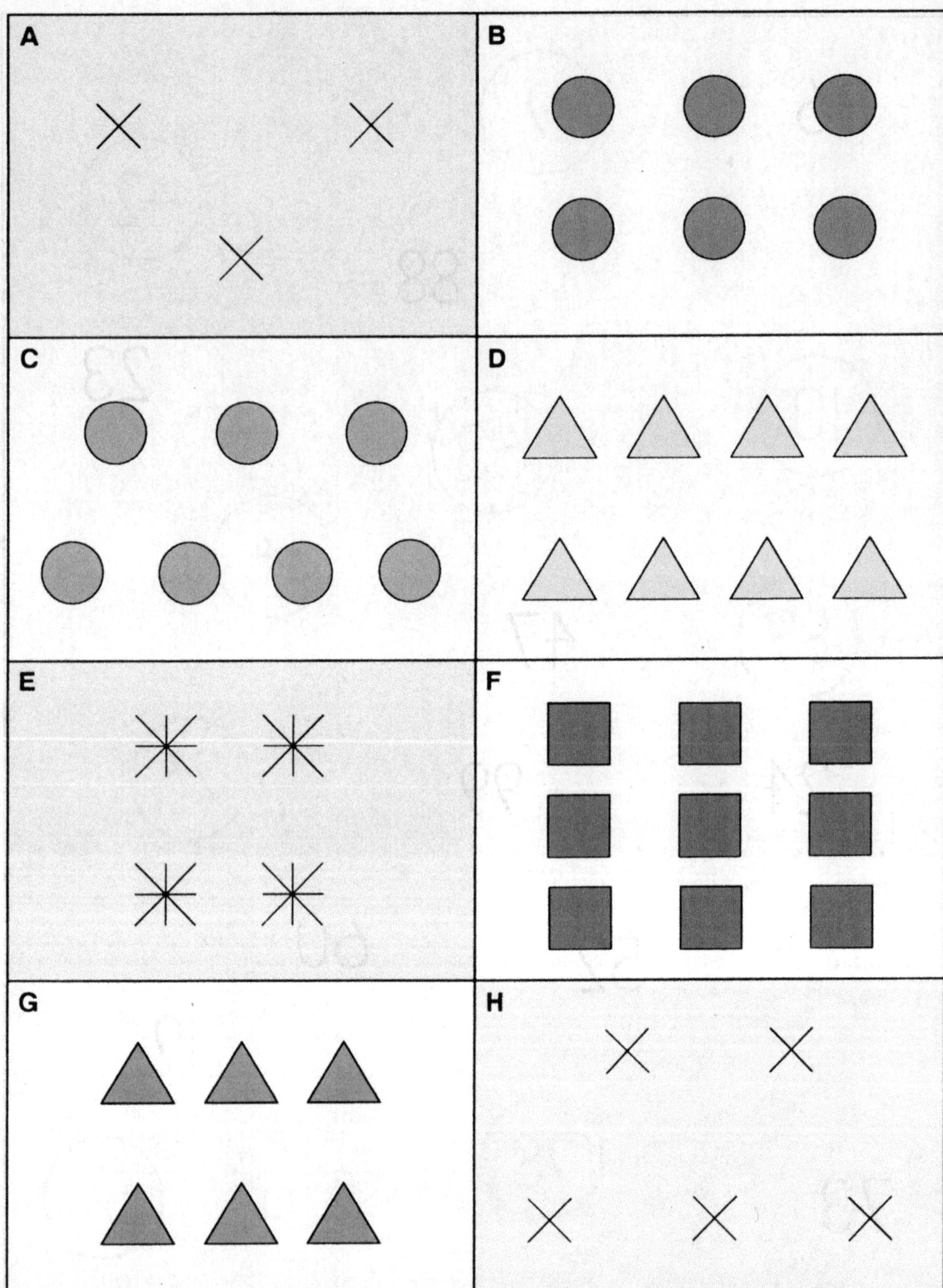
A
B
C
D
E
F
G
H

A
2t
1t
1t
5t
2t
2t
1t
1t
Use 2 coins.
B
5t
2t
1t
1t
1t
1t
1t
2t
2t
Use 3 coins.
C
5t
2t
1t
1t
1t
1t
2t
1t
2t
Use 4 coins.
D
1t
1t
1t
2t
2t
1t
1t
1t
1t
1t
2t
Use 7 coins.

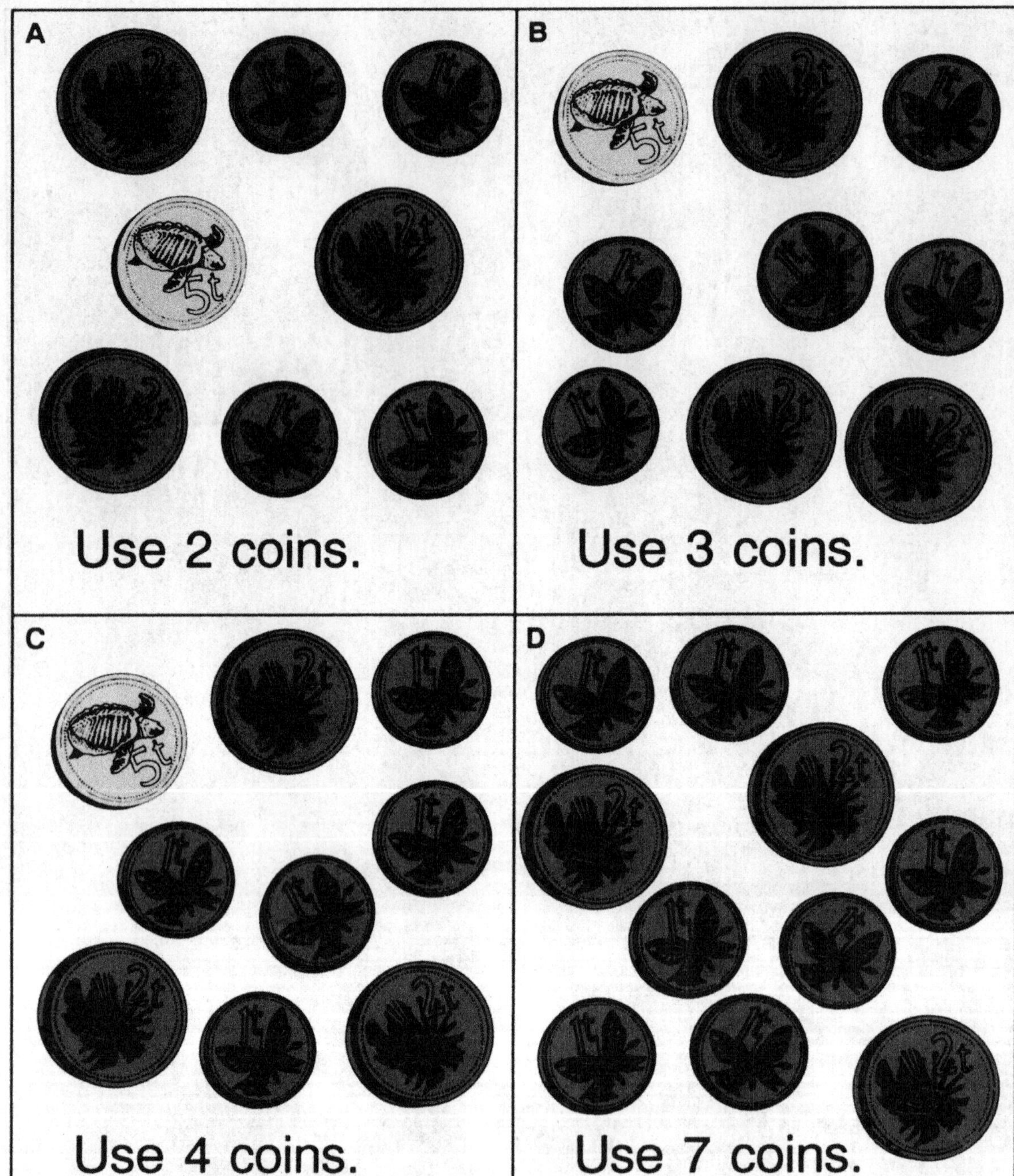

Make 7 toea.

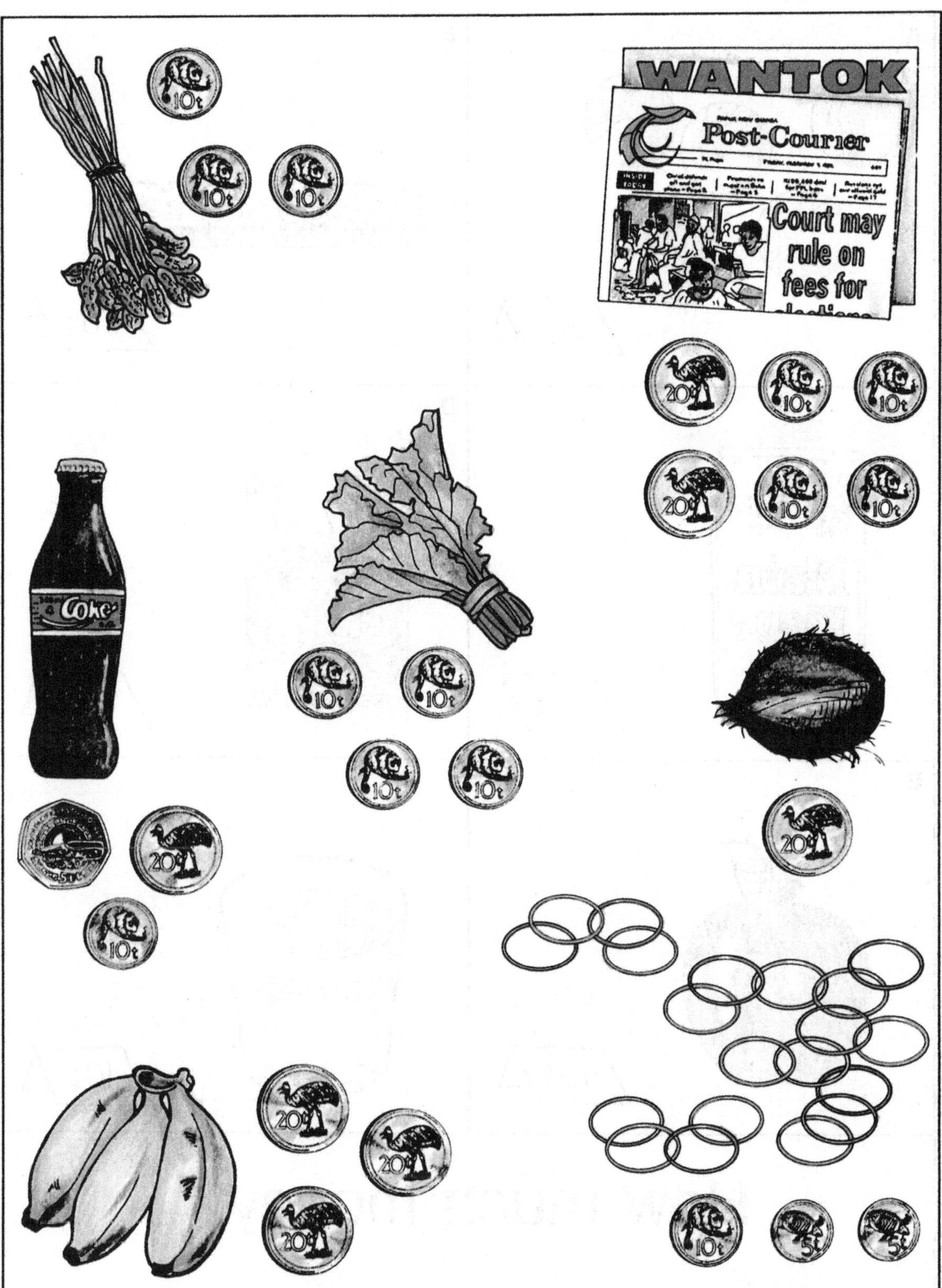
WANTOK
Post-Courier
Court may rule on fees for
Coke
10t
20t
5t

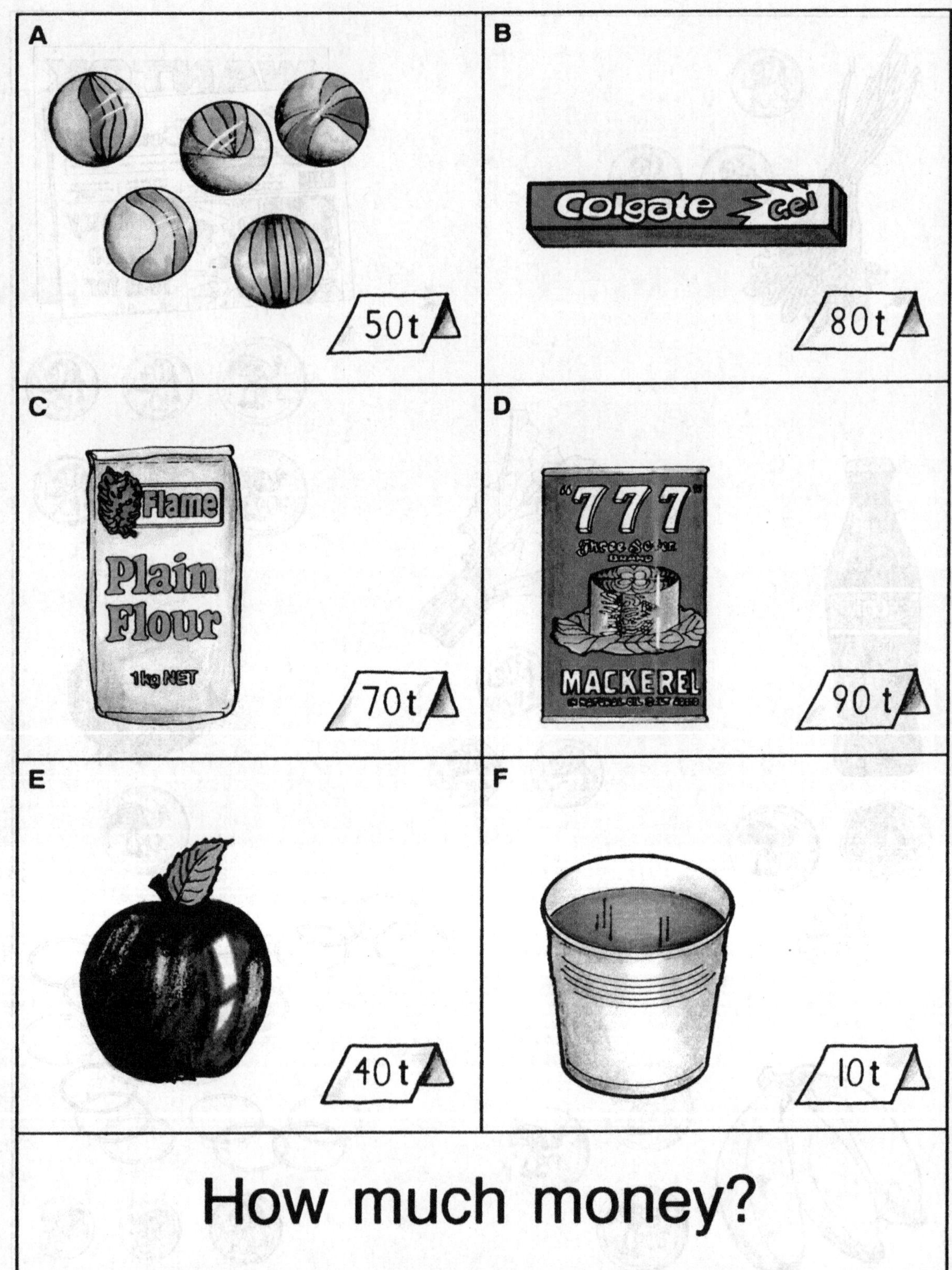

How much money?

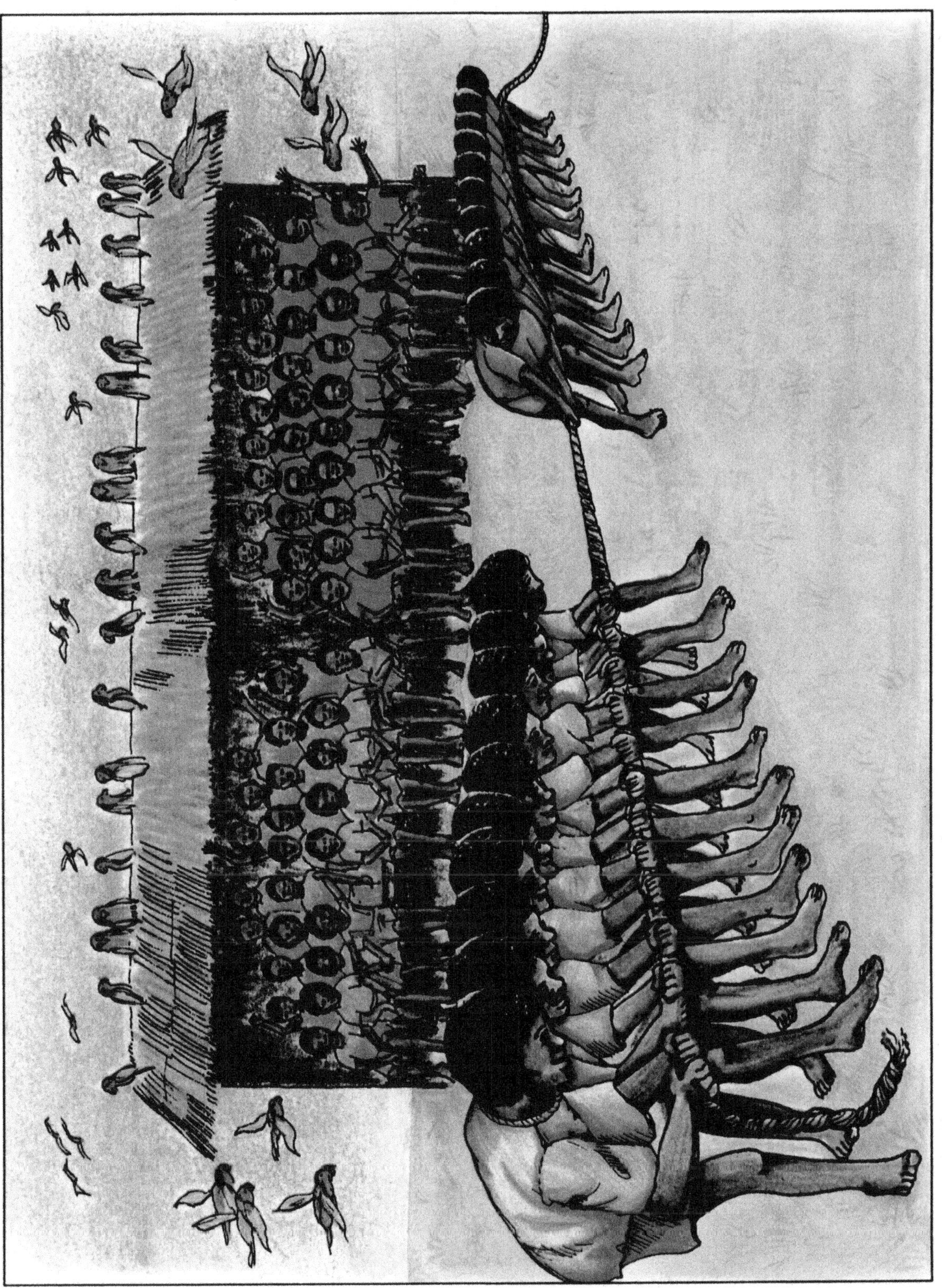

..AND WE ASSERT BY VIRTUE OF THE CONSTITUTION THAT ALL POWER BELONGS TO
PEOPLE – ACTING THROUGH THEIR ELECTED REPRESENTATIVES..
PARLIAMENT MAY MAKE LAWS HAVING EFFECT WITHIN AND WITHOUT THE COUNTRY FOR THE PEACE
ORDER AND GOOD GOVERNMENT OF PAPUA NEW GUINEA AND THE WELFARE OF THE PEOPLE

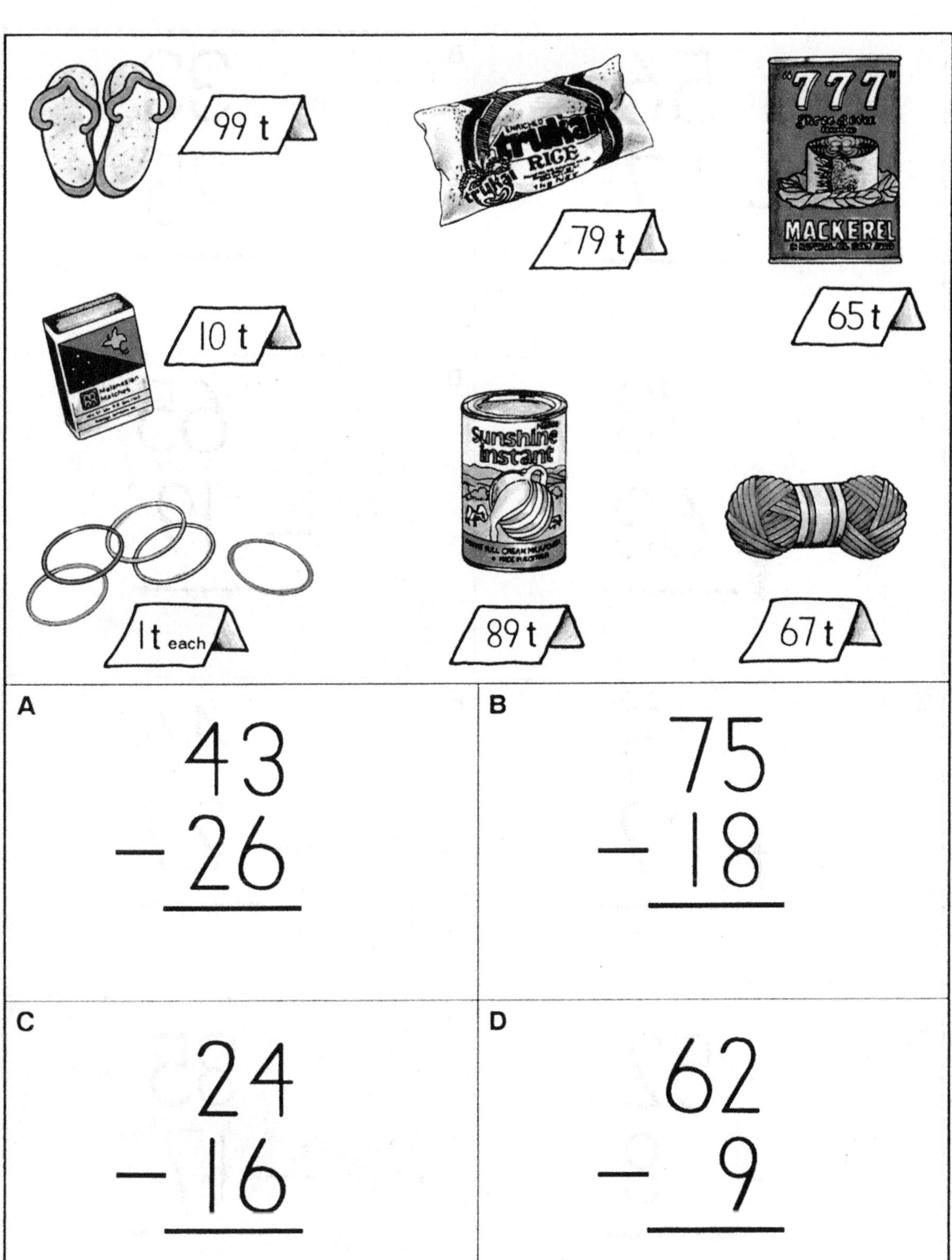
99 t
ENRICHED
trukai
RICE
79 t
"777"
MACKEREL
65 t
10 t
Melanesian Matches
Sunshine Instant
1 t each
89 t
67 t
A
43
− 26
B
75
− 18
C
24
− 16
D
62
− 9

A $\begin{array}{r} 56 \\ -\ 17 \\ \hline \end{array}$	**B** $\begin{array}{r} 33 \\ -\ 25 \\ \hline \end{array}$
C $\begin{array}{r} 78 \\ -\ 63 \\ \hline \end{array}$	**D** $\begin{array}{r} 65 \\ -\ 19 \\ \hline \end{array}$
E $\begin{array}{r} 15 \\ -\ 9 \\ \hline \end{array}$	**F** $\begin{array}{r} 46 \\ -\ 27 \\ \hline \end{array}$
G $\begin{array}{r} 52 \\ -\ 8 \\ \hline \end{array}$	**H** $\begin{array}{r} 35 \\ -\ 17 \\ \hline \end{array}$

19 18 40 15

6 35 22

25

9

28

65	44	24	46	38
31	25	53	68	63
43	37	62	34	55
57	49	15	54	50
60	59	33	75	28

$\begin{array}{r} 59 \\ +\ 24 \\ \hline \end{array}$	$\begin{array}{r} 21 \\ +\ 67 \\ \hline \end{array}$
$\begin{array}{r} 73 \\ -\ 19 \\ \hline \end{array}$	$\begin{array}{r} 18 \\ +\ 41 \\ \hline \end{array}$
$\begin{array}{r} 56 \\ -\ 34 \\ \hline \end{array}$	$\begin{array}{r} 40 \\ -\ 22 \\ \hline \end{array}$

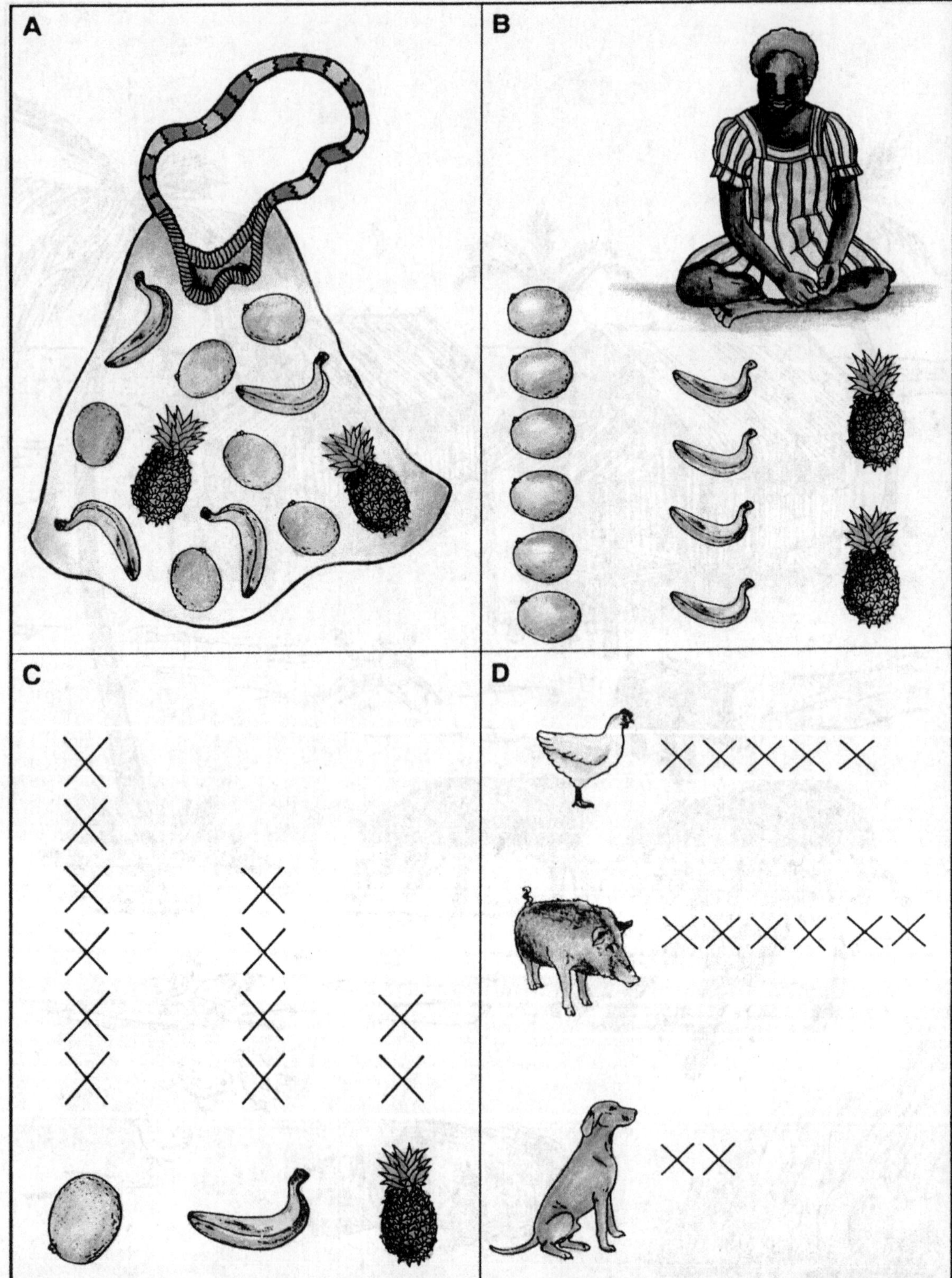
A
B
C
D

A
B
C

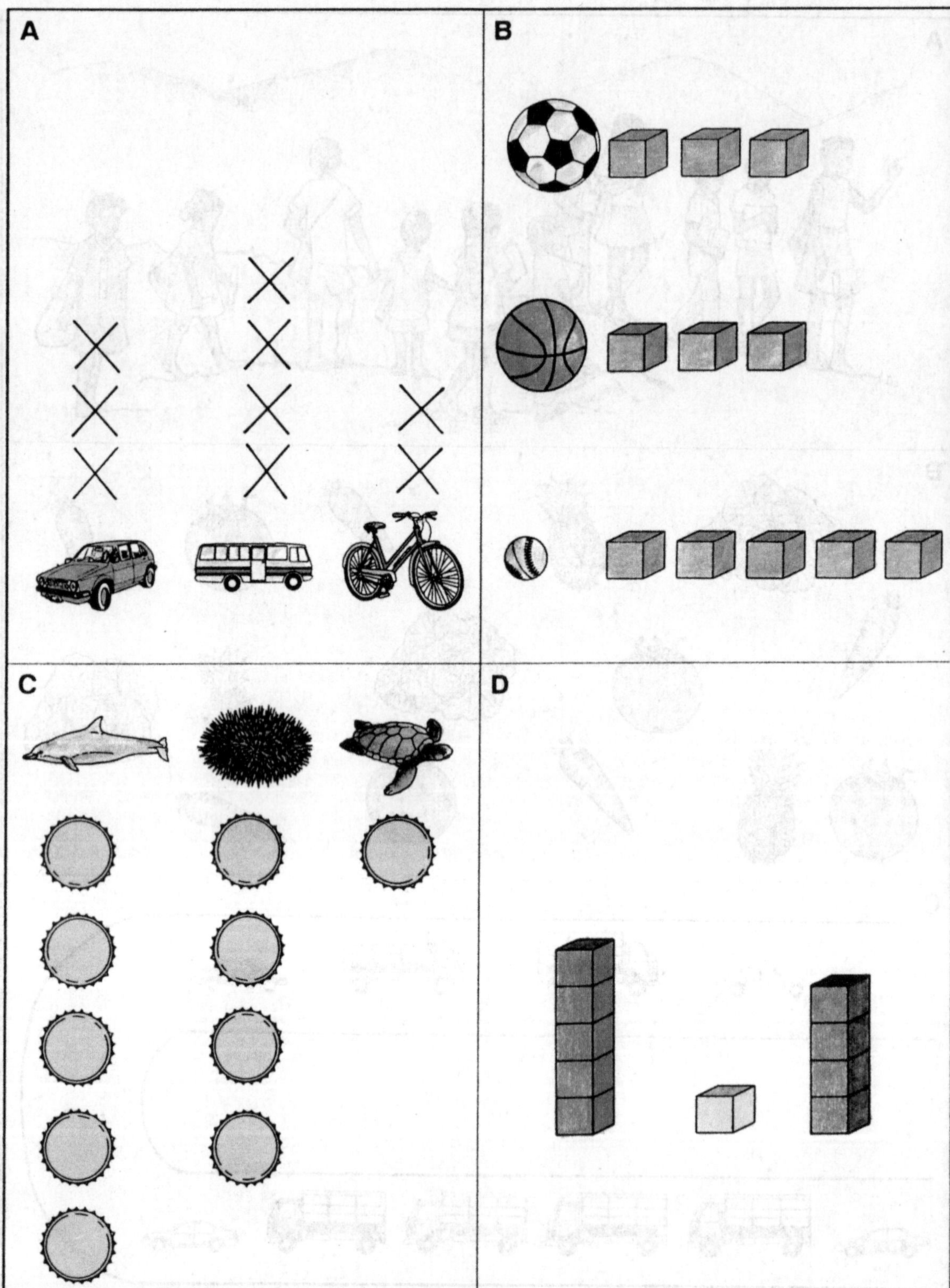
A
B
C
D

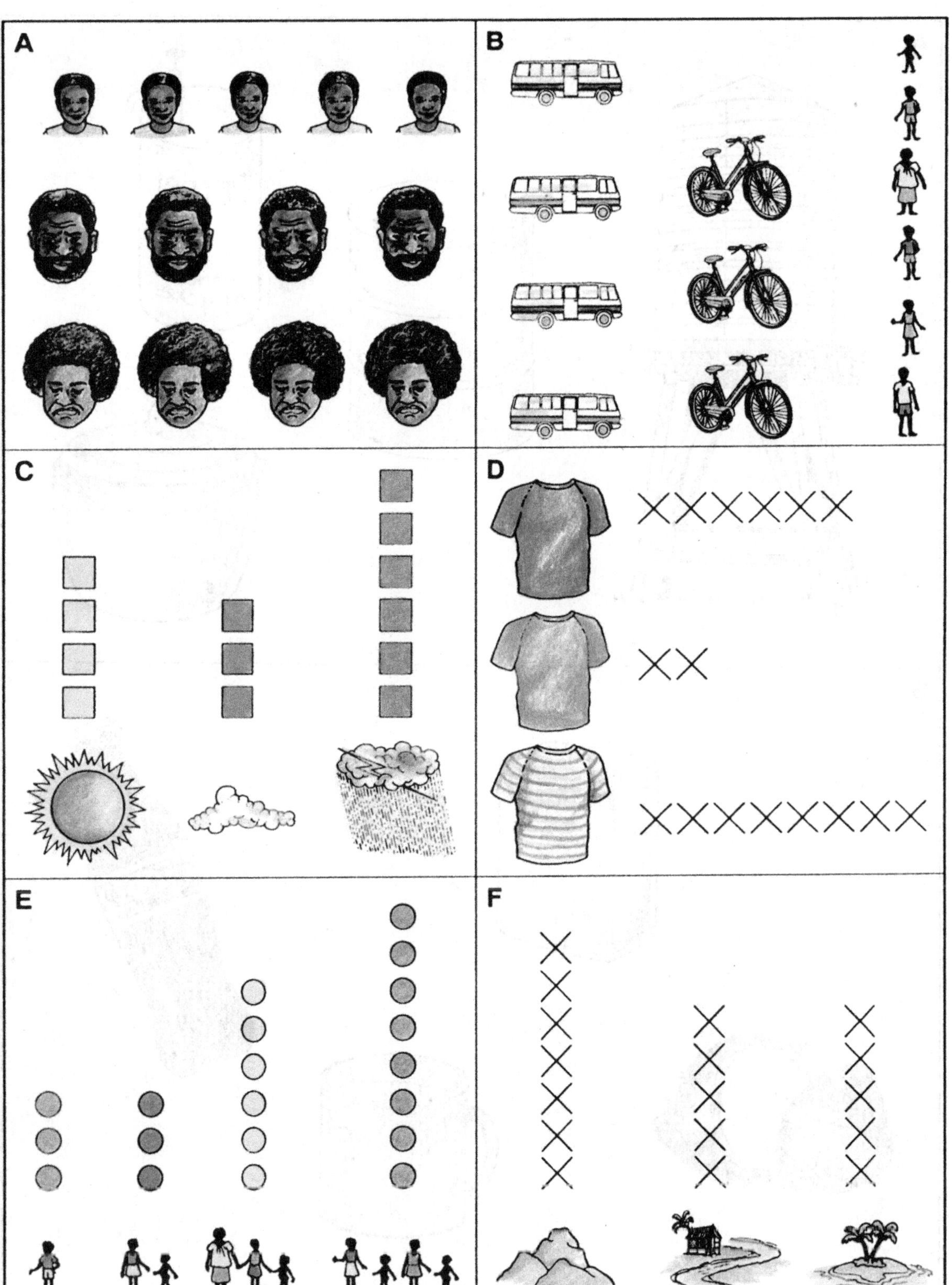
A
B
C
D
E
F

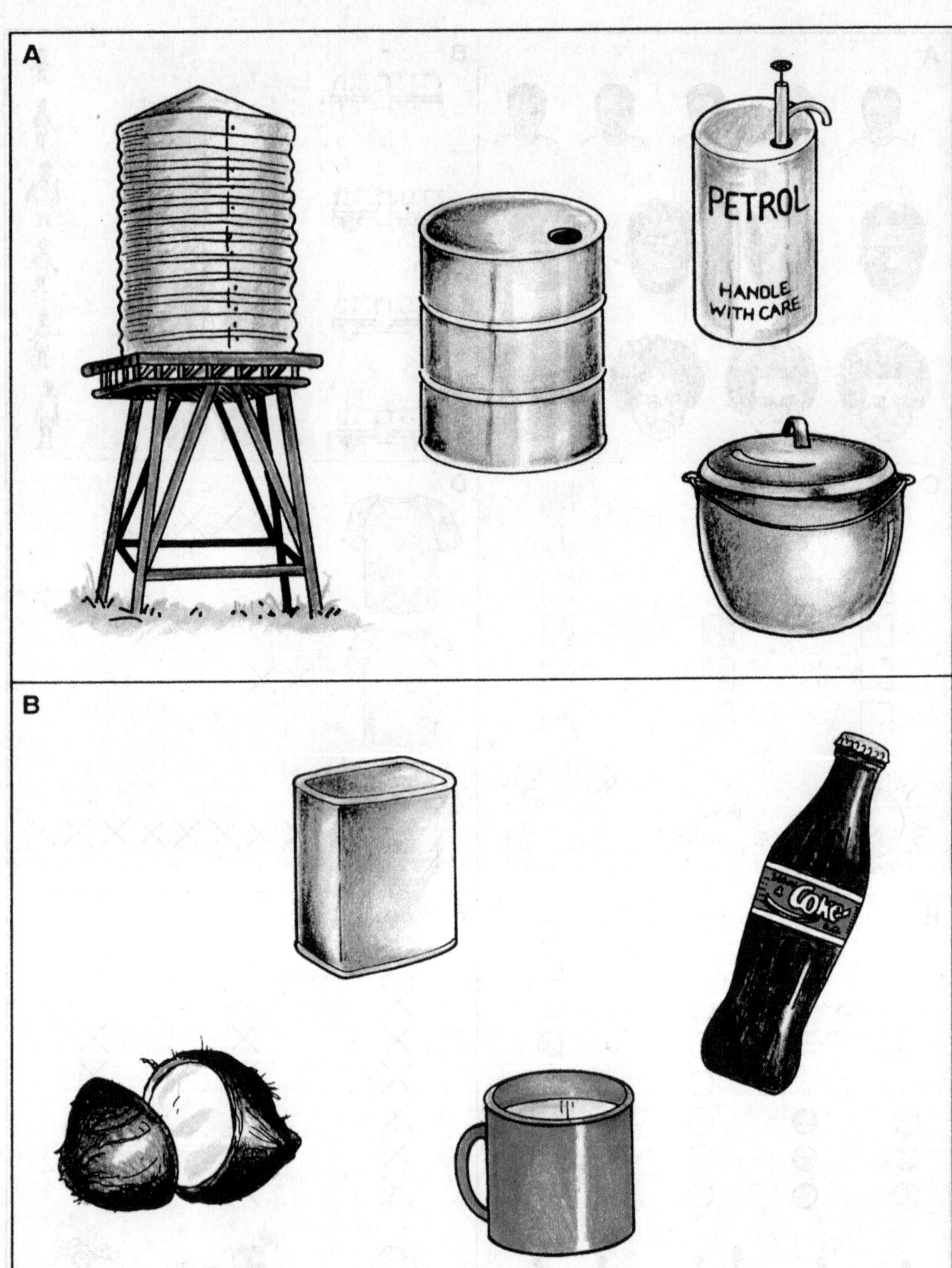
A
PETROL
HANDLE
WITH CARE
B
Coke

Coke
Melanesian
Matches

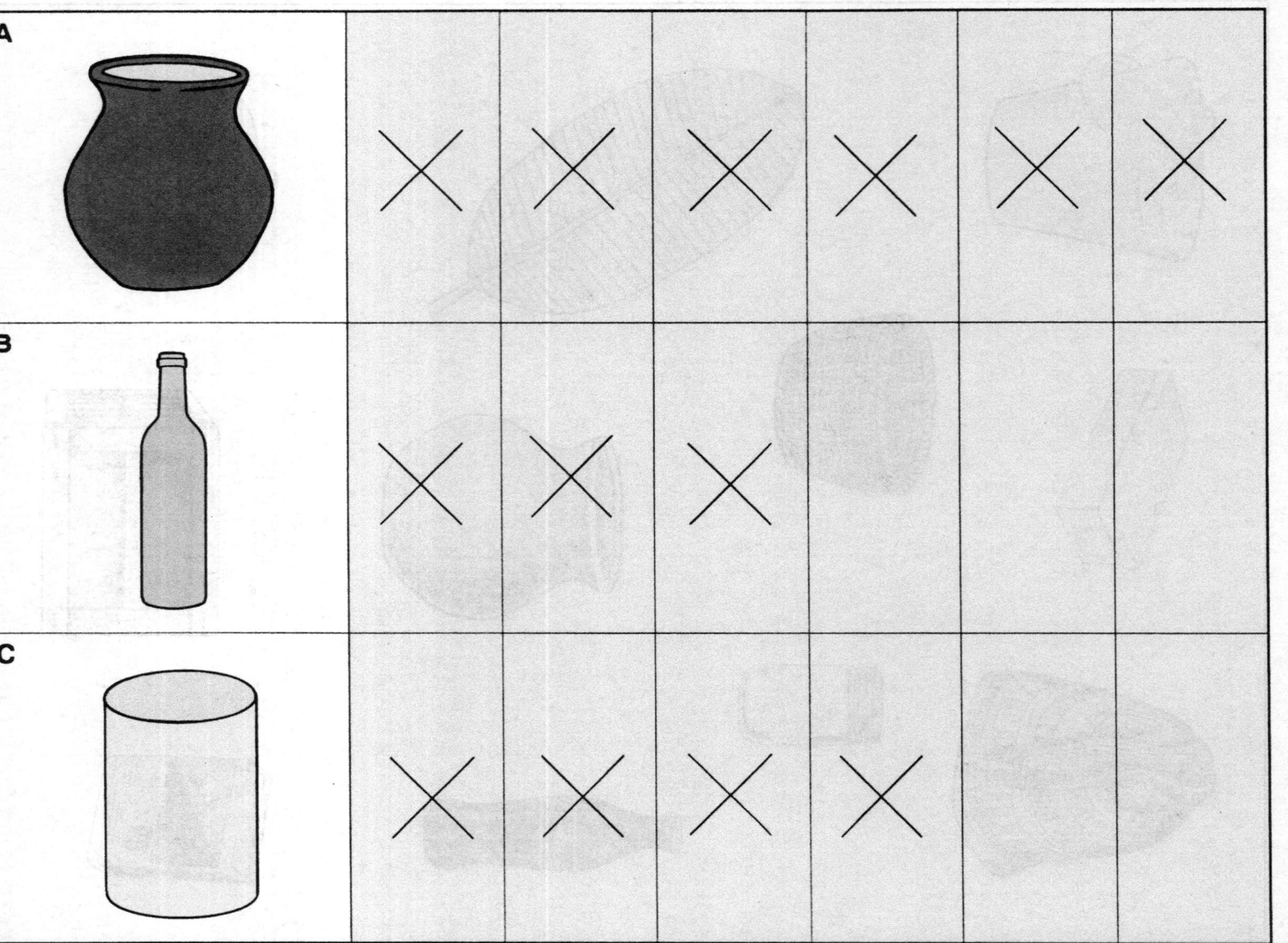
A
B
C

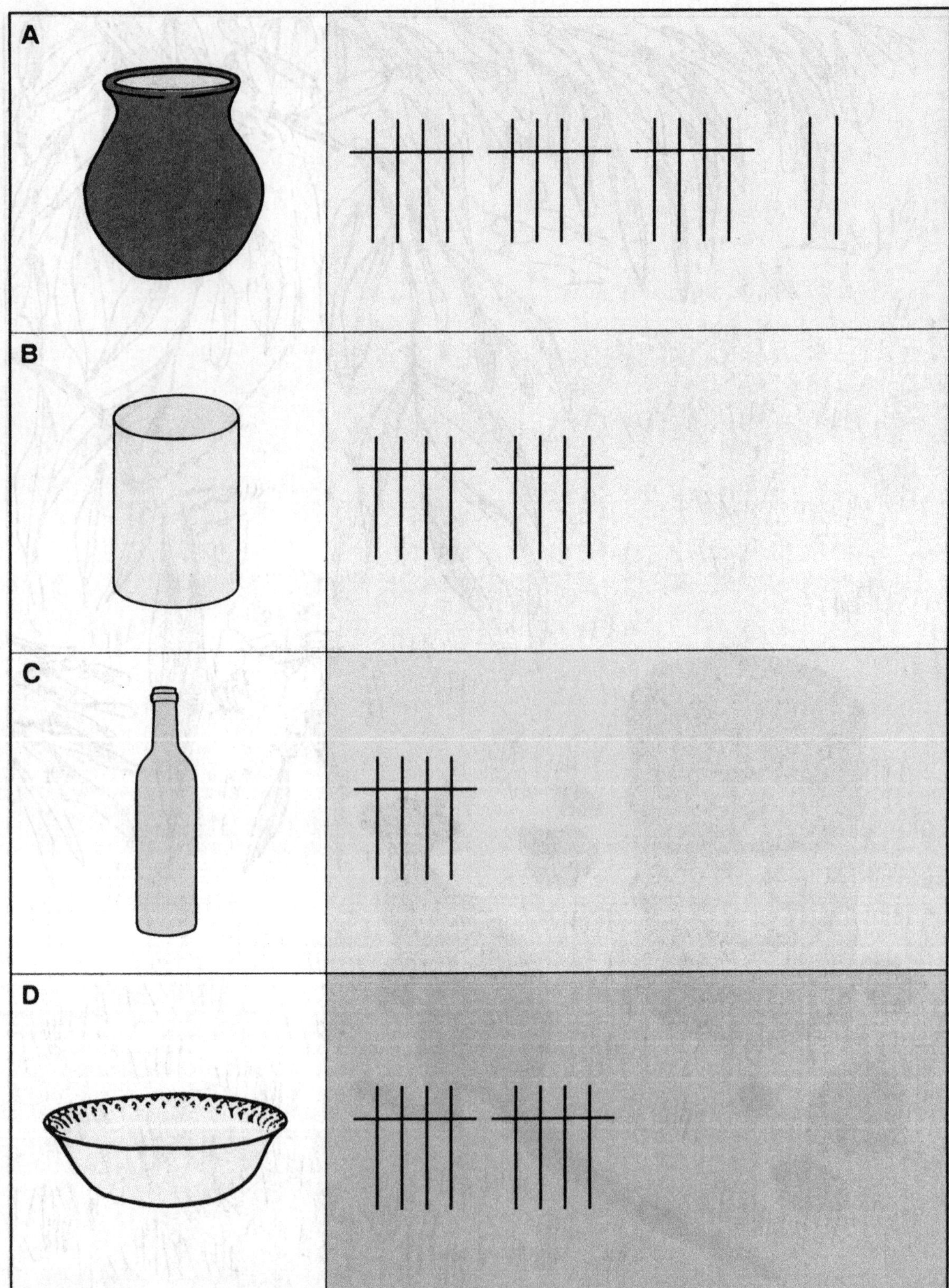
A
B
C
D

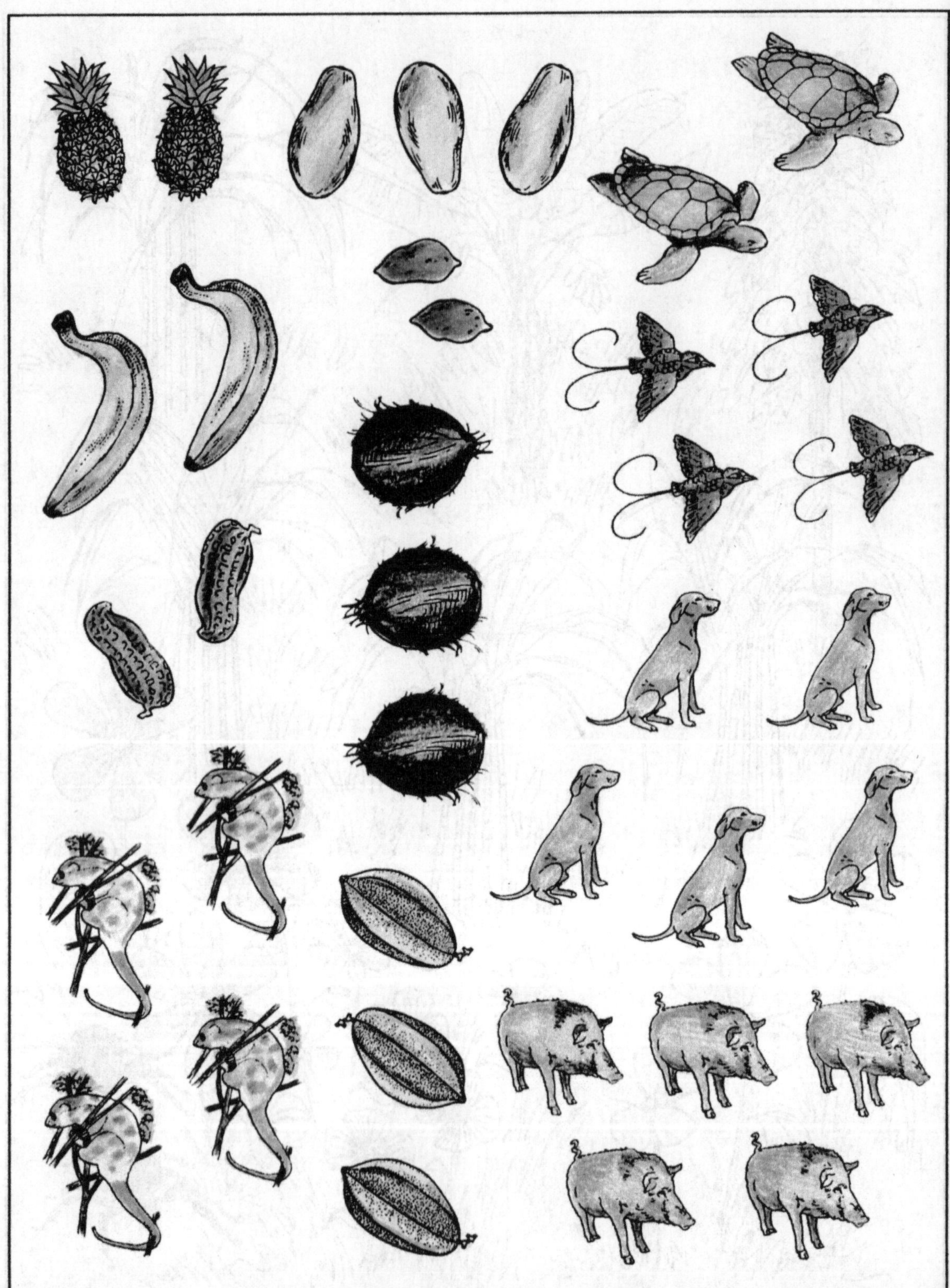

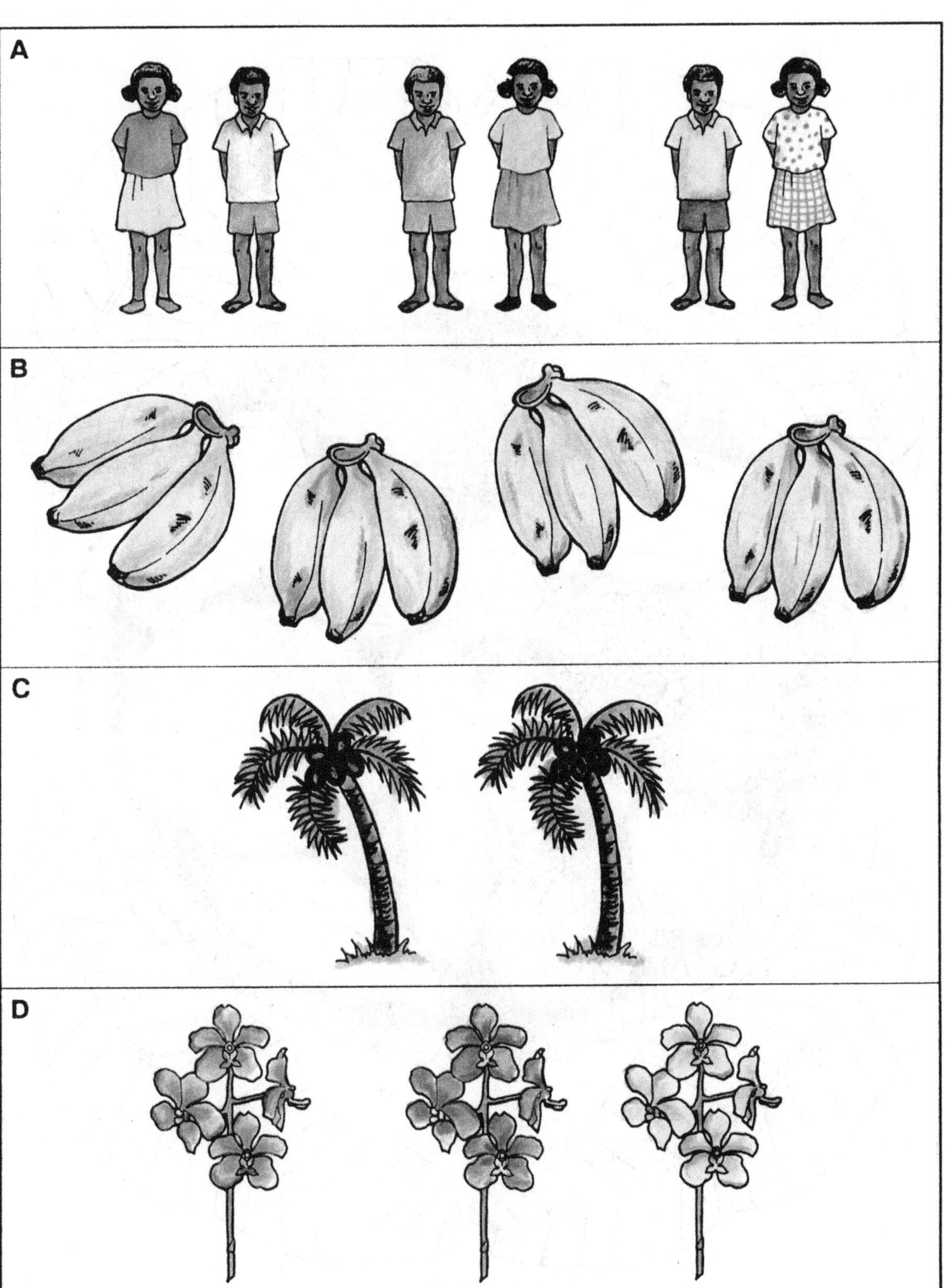
A
B
C
D

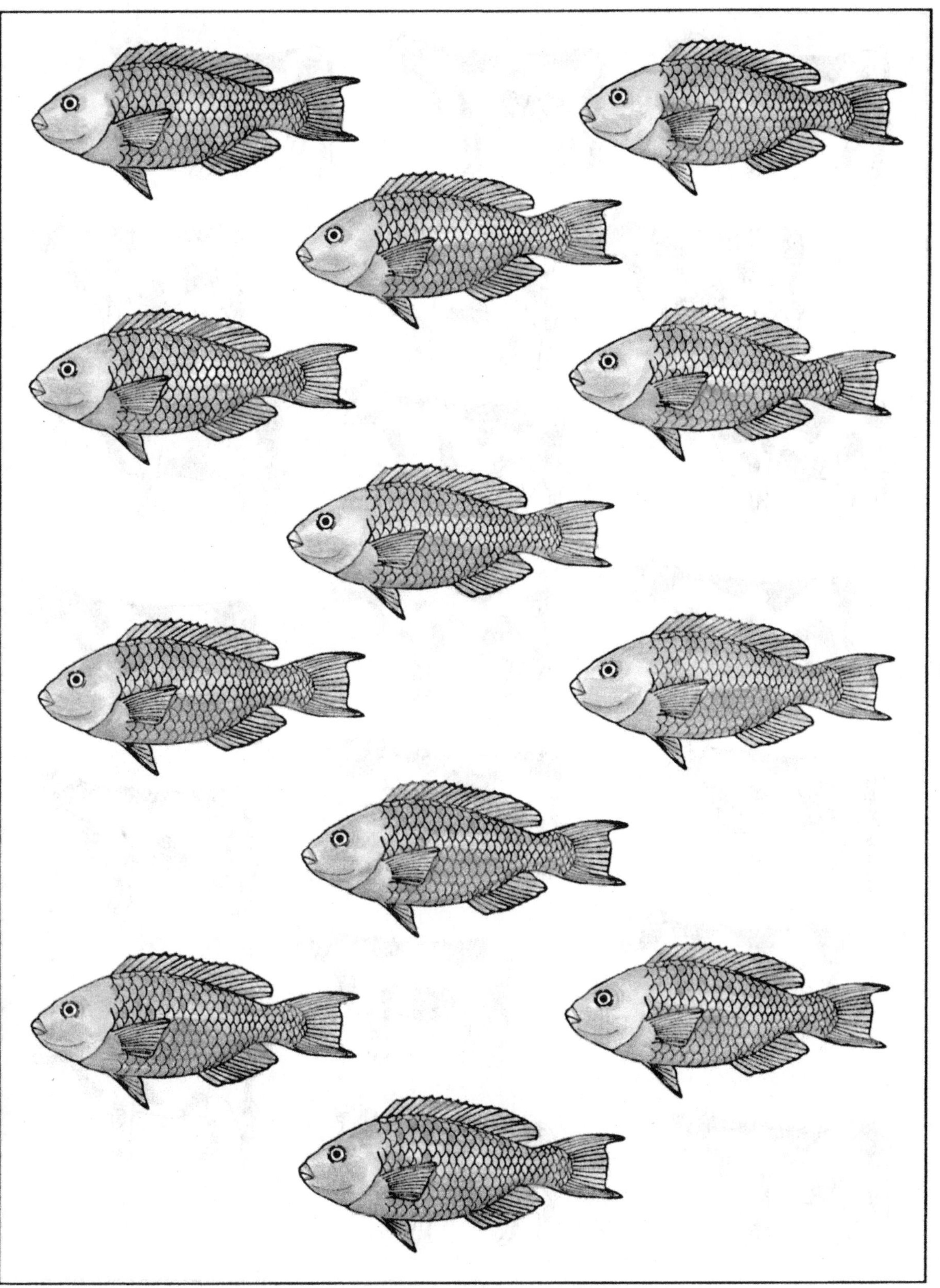

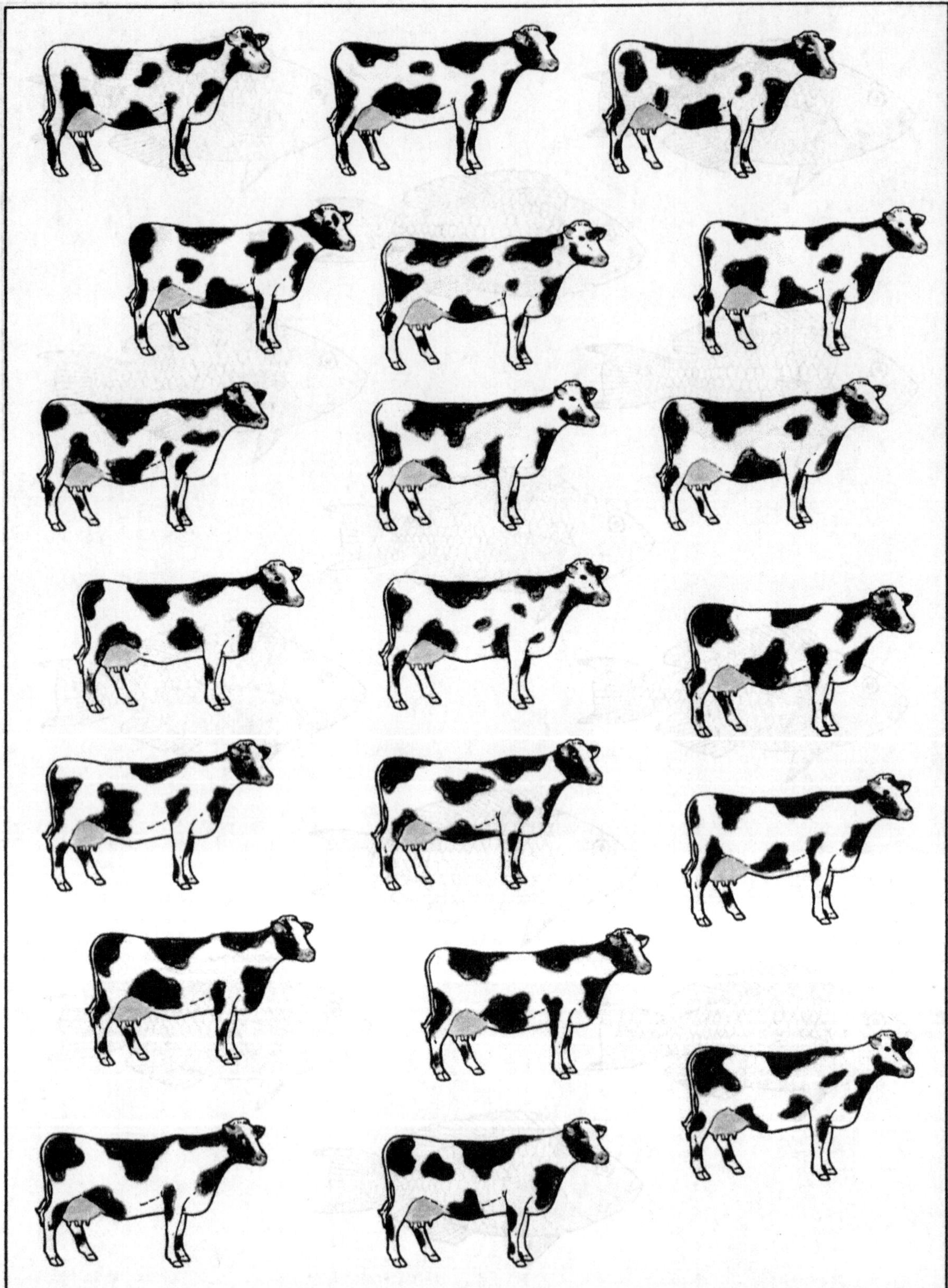

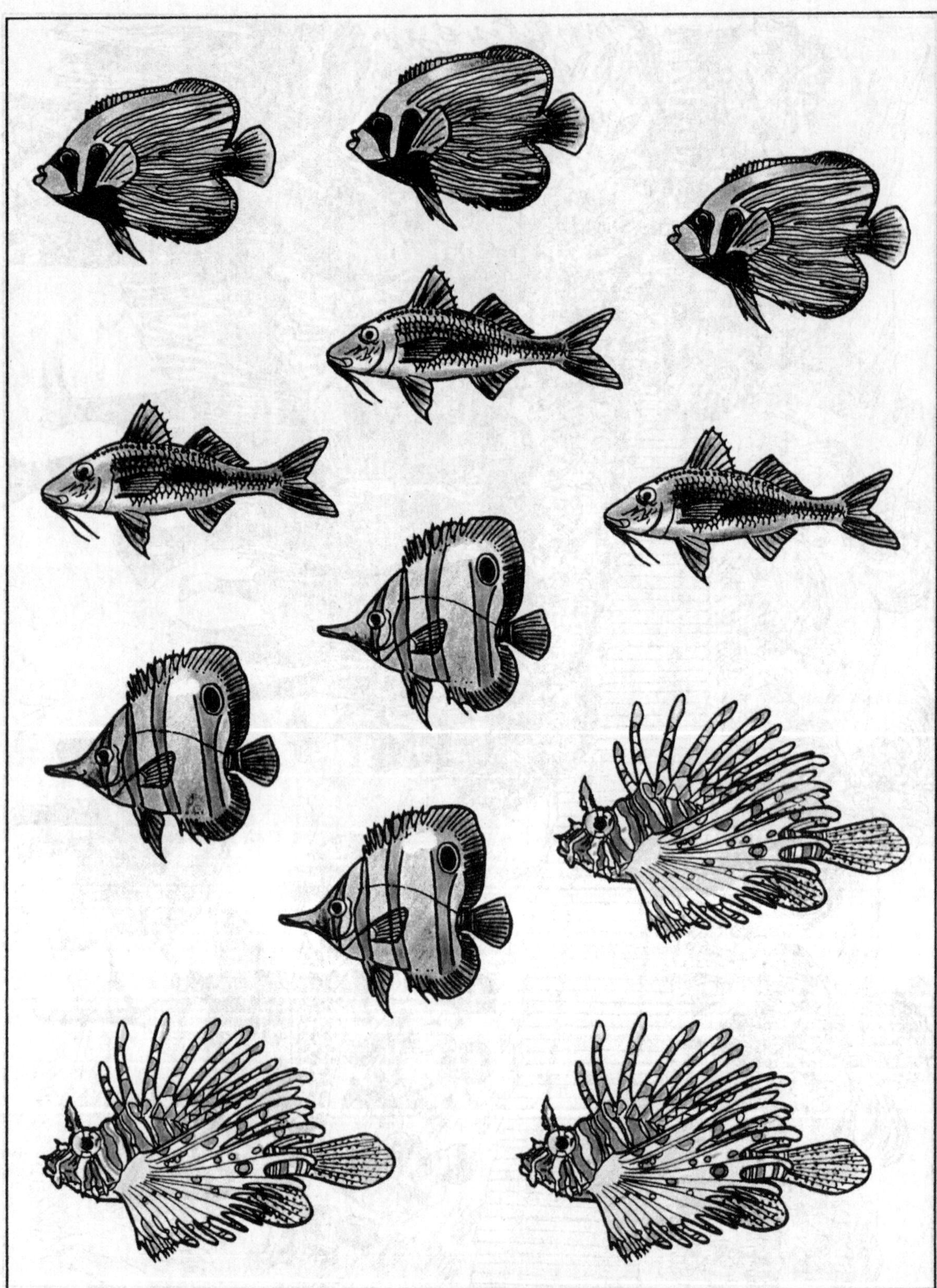

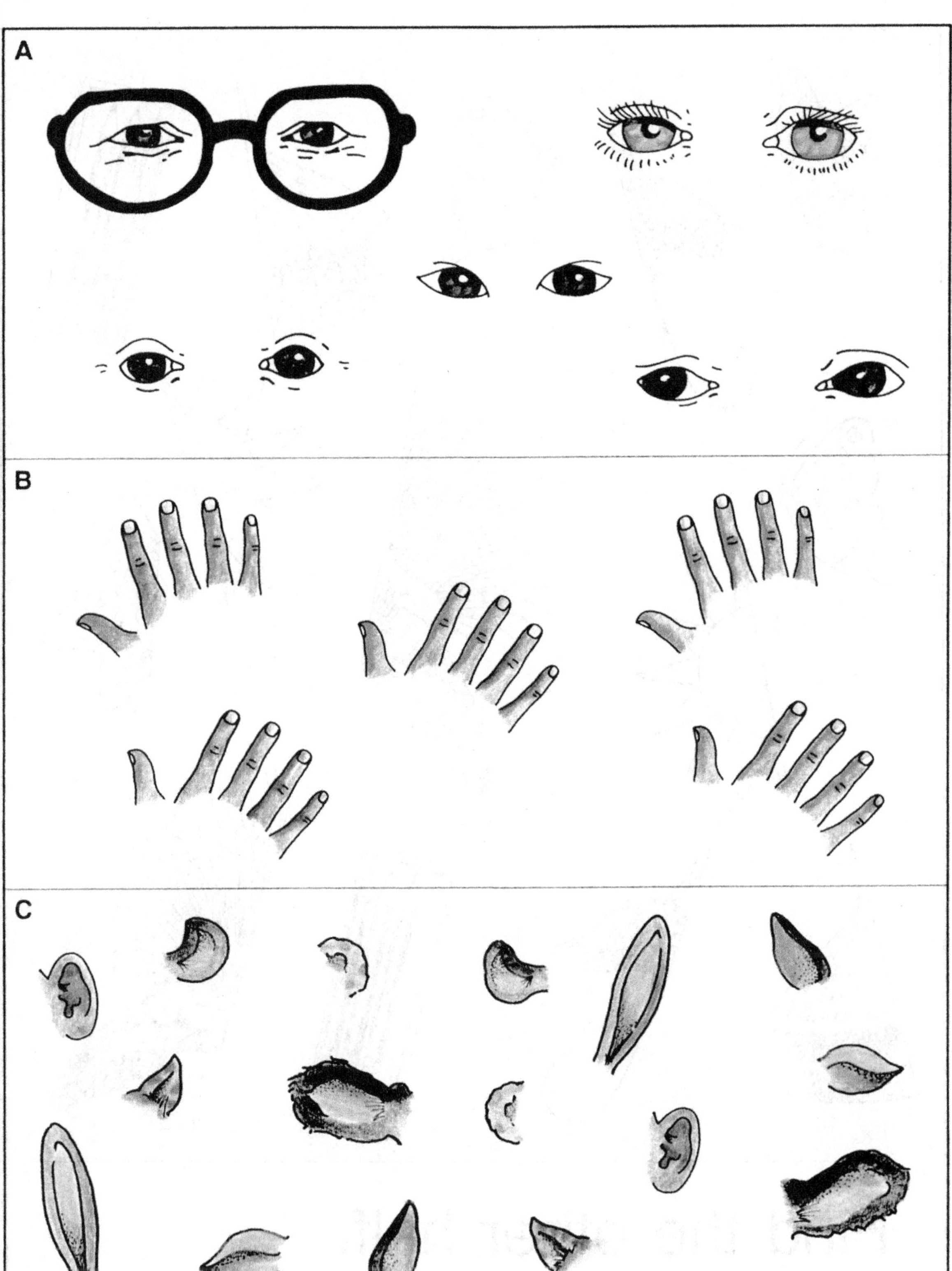
A
B
C

Find the other half.